THE ILLUSTRATED GUIDE TO
CHICKENS

How to Choose Them, How to Keep Them

Skyhorse Publishing

Skyhorse Publishing books may be purchased in bulk at special discounts for sales promotion, corporate gifts, fund-raising, or educational purposes. Special editions can also be created to specifications. For details, contact the Special Sales Department, Skyhorse Publishing, 307 West 36th Street, 11th Floor, New York, NY 10018 or info@skyhorsepublishing.com.

Skyhorse® and Skyhorse Publishing® are registered trademarks of Skyhorse Publishing, Inc.®, a Delaware corporation.

www.skyhorsepublishing.com

10 9 8 7 6 5 4 3 2 1

Library of Congress Cataloging-in-Publication Data is available on file.

ISBN: 978-1-61608-425-7

Printed in China

The author and Skyhorse Publishing, Inc. have made every effort to ensure that all advice in this book is accurate and safe, and therefore cannot accept any liability for any resulting injury, damage, or loss to persons or property, however it may arise.

Contents

Buff Orpington
hen with chicks

As Patron of both the Rare Breeds Survival Trust and the Poultry Club of Great Britain, I could not be more pleased to have been invited to contribute the foreword to this beautiful book, *An Illustrated Guide to Chickens.*

My family's interest in poultry goes back to my Great Great Great Grandmother, Queen Victoria, who was presented with a flock of the first Brahmas ever seen in this country. I understand their great size caused quite a stir! My Grandmother, Queen Elizabeth The Queen Mother, kept Buff Orpingtons and was enormously proud of her Patronage of the Buff Orpington Society. In my own case, ever since I was a child and used to collect eggs from the farm at Windsor, I have had an interest in chickens and have continued the family tradition at Highgrove with both Marans and Welsummers.

It is a mark of the times in which we live that some fifty breeds of chicken are now endangered and I hope and pray that this book, which is filled with practical information on keeping birds, will draw attention to their plight and encourage existing and new breeders to play their part in rebuilding their numbers.

Introduction

Which came first, the chicken or the egg? Science has now come up with an answer to this riddle. It seems that all new species develop from a genetic mutation; if this is successful for survival then the new genes are passed on to following generations and a new species is born. The first chicken was a mutation of its avian parents and its life began in the egg before it hatched. So it was the egg that came first and hatched into what eventually became the Red Junglefowl or *Gallus gallus* of Southeast Asia, one of four distinct wildfowl, which is thought to be the ancestor of all domestic fowl.

There are historical references to some kind of domestic fowl as far back as 3000 BC, and by 1400 BC the Chinese and Egyptians had invented crude incubators from clay that hatched vast numbers of chicks at a time. Alexander the Great is credited with introducing chickens to Europe around 500 BC and the Romans continued to spread them far and wide. At first they were raised more for cockfighting than anything else, a sport that was popular worldwide.

Cockfighting is a fight between two cockerels whose natural aggression causes them to fight to the death, either using their own natural spurs or, more commonly, with razor-sharp blades known as cockspurs or gaffs attached to their legs. The fight only ends when one bird is killed or is too tired to continue, and in many cases the victor is so severely injured that it dies as well. Cockfighting is now illegal in the USA, the UK and most of Europe.

Eventually poultry reached Britain, probably introduced by the Romans, though it is possible they had already arrived by other means, and were eventually introduced to the New World. The conquistadors found poultry already established in South America, probably brought by Polynesian traders when they arrived in the early 15th century. They soon spread north and were added to by the French settlers in Canada and colonists from England.

Nowadays there are literally hundreds of breeds of chickens all over the world. The following pages will give you an insight and hopefully help you make a selection.

Chickens are a joy to keep; they are adaptable creatures and whether you are lucky enough to be able to let them free range or have to keep them in a coop, there will be a breed to suit you. There can be little that compares with the sight and sound of a magnificent cockerel or mother hen with her 12 little troopers following obediently in her wake. They will eat up all your leftovers and turn them into delicious, nourishing eggs—finding your first new-laid one will be a thrill and collecting them daily will remain a pleasure.

Sadly many of the older breeds are now in danger of extinction, but some stock does still remain and it is within anyone's power to help preserve them. The rarer the breed the more you will be able to ask for any surplus stock if you wish to sell some, but equally they may be difficult to rear—which, indeed, is why they are suffering in the first place.

Choosing the right chickens

Having decided hens are for you, your next decisions will be hybrids or pure bred, large fowl or bantam? Do you want eggs or meat? Are they for show? Nearly all breeds of hen have a bantam equivalent; there are also several breeds that are pure bantam with no large version. Bantams make delightful pets and can become very tame—and they obviously need less space than their larger cousins. Most breeds will go broody frequently, making very good mothers, but they do lay very small eggs.

Adult hen   Pullet   Bantam

A hybrid is a cross of two or more breeds that have been carefully selected to produce birds with prolific egg-laying tendencies. There are many well-known hybrids and there will be a breeder near you that stocks one or more. A hybrid will be the most economic egg-producing machine, so if you don't want chicks or a cockerel then these are for you as they very rarely go broody and would not breed true.

Pure breeds won't lay quite as many eggs as hybrids and some will tend to go broody, but you will be able to breed from them. It can be very rewarding to keep a rare breed and help preserve a species. Pure breeds come in an amazing array of colors, characters, shapes and sizes, some being better layers than others, some producing better carcasses, and some just looking extraordinary.

Yamato-Gunkei

Thüringian

Transylvanian Naked Neck

Pure breeds are broken down into soft feather and hard feather. The soft feather again comes in large, small, and bantam—these are predominantly layers, though some are termed dual purpose if they produce a good meaty carcass as well. The Mediterranean breeds, such as Ancona or Leghorn, tend not to go broody, have white earlobes, and lay good numbers of white eggs. Asian soft feather are characteristically large with fluffy feathers and feathery legs, such as Brahma, Cochin, and Langshan. The Asian gamefowl all have hard, tight feathering, are aggressive (as they were developed to fight), and tend to go broody frequently while laying few eggs—this is not to say they are unpopular, as even though cockfighting has been banned since 1849, there is plenty of competition in showing and these birds have strong and charming characters that endear them to their owners. It must be borne in mind, however, that they have been bred as fighting birds and this they will do. They can only be kept in pairs or trios and no new birds can be introduced as they will simply not be accepted.

How the book works

Over the following pages, 100 breeds of chicken are described, with illustrations. All individual breeds have different character strains within them, so please note that the descriptions given are a guide only and where a hen is described perhaps as being flighty, this will be the general characteristic of that breed; certain lines of the same breed may be calm, but they will be the minority.

The illustrations are typical types of each breed, but if you are thinking of showing your birds, then you should consult your country's breed standard (see below).

TYPE	CLASS	ORIGIN	EGG COLOR	STATUS
Layer	Soft feather light	Europe	White	Common
COMB TYPE	FEATHER COLORING	BROODY	NO. OF EGGS/YEAR	BREED STANDARD
Rose	Gold penciled Silver spangled White Black	Not often	Prolific	APA PCGB Europe

TYPE: whether the breed is suitable for laying eggs—**layer**—or will lay few eggs but produce a fine carcass—**table**. It may lay plenty of eggs and also produce a good carcass, known as "dual purpose", or could be purely **ornamental**.

CLASS: game birds and Asian gamefowl are classed as **hard feather**, which means they have short feathers so tight to their bodies that in some cases the skin shows through. **Soft feather**, where the plumage is looser and fluffier, comes in two sizes, light and heavy—**light** being mainly birds of Mediterranean origin that are excellent layers, and **heavy** those larger breeds that are frequently dual purpose. **True bantams** are birds that have no large equivalent—most large breeds having a bantam equivalent.

ORIGIN: where the breed first originated.

EGG COLOR: which color or colors of egg are produced by the breed.

STATUS: whether the breed is common, fairly common, or rare. In this context, rare means probably fewer than 500 breeding females in existence—in some cases considerably less than that.

COMB TYPE: breed found with one or more of the six types of comb.

FEATHER COLORING: each breed may come in several different colorways.

BROODY: whether the breed tends to go broody frequently, occasionally, or not at all.

NO. OF EGGS/YEAR: a loose guide to give an idea of how many eggs to expect each year. N.B. this is dependent on many things, such as food, comfort, age, etc.

0–50	very few
50–100	few
100–150	moderate
150–200	prolific
200 or more	very prolific

BREED STANDARD: whether the breed is recognised by a particular poultry organization and has a standard, meaning that a set of detailed guidelines are laid down with exacting details of plumage, head, and leg coloring, etc., with a scale of points for judging and serious defects.

APA: American Poultry Association
PCGB: Poultry Club of Great Britain
Europe: Various European countries have a standard for this breed

Roosters or cockerels

Blue laced
Wyandotte rooster

You do not need a rooster (known as a cockerel until his first molt) for your hens to lay eggs; this they will do anyway, but without one they will not be fertile.

There are definite pros and cons to keeping a rooster. On the pro side, you will be able to hatch chicks, he will look magnificent, and to an extent he will protect his flock from predators, or at least be the first line of defense. He will also be charming to the hens, calling them over when he finds some food—he has an ulterior motive, of course, but the hens always respond. On the con side, he will crow. He will crow frequently during the day and start before dawn—various bantam roosters even crow in the middle of the night. To some, this will be a delightful sound of the country, but to others it will be an unpleasant nuisance, so this should be an important consideration. You can keep him quieter by making sure he stays in a dark henhouse until a reasonable hour; he will also find it difficult to crow if he can't lift up his head, so a high perch or cramped conditions may keep him quiet to an extent.

Another con is that if you only have three or four hens his favorite one may suffer from his attentions and develop a bare back and neck—if you are wanting to show your birds, you can acquire a "saddle", which is a sort of light cover that fits on the hen's back. A rooster can service at least ten hens, so keeping six or more should spread his attentions.

Finally, he may very well be aggressive. He will certainly be aggressive toward another cockerel, but he may also find you a threat and take, literally, to attacking the hand that feeds him. Plenty of roosters are perfectly friendly, particularly the larger breeds, and even gamefowl tend only to be aggressive toward their own kind.

Don't forget, he will be another mouth to feed with no return—unless you are considering coq au vin.

Game rooster

Parts of a chicken

Comb

Eye

Beak

Wattle

Neck hackles

Ear

Earlobe

Tail covert

Main tail

saddle hackles

Keel

Wing

Thigh

spur

side hangers

Tail sickle

Combs

The comb is the red fleshy growth on the top of a chicken's head and is usually larger in the cockerel. It is one of the distinguishing features of each breed and comes in many varieties. Chickens cannot sweat and so blood pumping through the exposed comb and wattles naturally cools the rest of the body. It is also a useful signal of the overall health of the bird and in the hen indicates whether or not she is in lay.

In lay   Not in lay

Single: the most common type of comb. Single combs come in many sizes, may be upright or flop to one or both sides, and are usually larger in the cockerel. There may be different numbers of serrations or spikes.

single   Twisted single   single to one side

Rose: the second most common type found. Rose combs cover the top of the head like a flat cap with a tapering spike at the back and are covered in small round knobbles. The spike may follow the line of the neck, be horizontal, or turn up at the end.

Rose comb

Pea or triple: a low comb with three ridges, the middle one slightly higher than the other two and covered with small pea-like protuberances.

V-shaped: this is made up of two horn-like growths joined at the base.

V-shaped

Pea or triple

Walnut or cushion: a small comb with no spikes or protuberances—sits somewhat forward on the chicken's head.

Cup: this is really two single combs joined at the front and back and resembles a crown.

Walnut or cushion

Cup

Dubbing

The Old English, Modern, and certain other game birds can only be shown if the cockerels have been "dubbed". This is the removal of the comb and wattles when the birds are approximately six months old. Combs are trimmed close to the line of the head and the wattles similarly to the throat, although each breed has its particular style and breed standards should be consulted. This practice has now been banned in some European countries. Originally, dubbing was to prevent injury when the birds were used for fighting, but nowadays it is purely cosmetic. In cold climates, it does help to protect them from frostbite.

Blue red Modern
Game rooster

Feathers

· ·

Plumage

Chicken feathers come in an astonishing range of colors and patterns, which help to make each breed recognizable. The plumage plays an important role, protecting the chicken from rain, cold, and sun, and they must spend a considerable part of their time maintaining it. This is done by preening. Each feather has an axis or shaft, on to either side of which are fixed the vanes; each vane has barbs on either side, which cling together but need to be "combed" by the chicken, who also applies oil from a gland at the base of its tail. Dust bathing also plays an important part in feather maintenance.

Cockerels can be distinguished from hens by the fact that some of their feathers take on a different shape. Their hackle and saddle feathers are thinner and longer than a hen's and they also develop sickles, which are the spectacular curved feathers on either side of the tail. Some breeds have much fluffier feathers than others, and game breeds have very tight feathering that often leaves a strip of bare skin down the breast. There may be feathering on the legs, and some breeds sport beards, muffs, and crests.

Every year, hens molt, generally at the beginning of fall, and replace their old feathers with new. As feathers are largely made up of protein, this takes a good deal of the hen's energy, and it is important to give her plenty of replacement protein in the form of good-quality layers' ration at this time. She will stop laying until her molt is complete, which could take anywhere between six and twelve weeks, and if the days are growing shorter, she may not start laying again until they start to lengthen after the winter solstice.

Types of feather

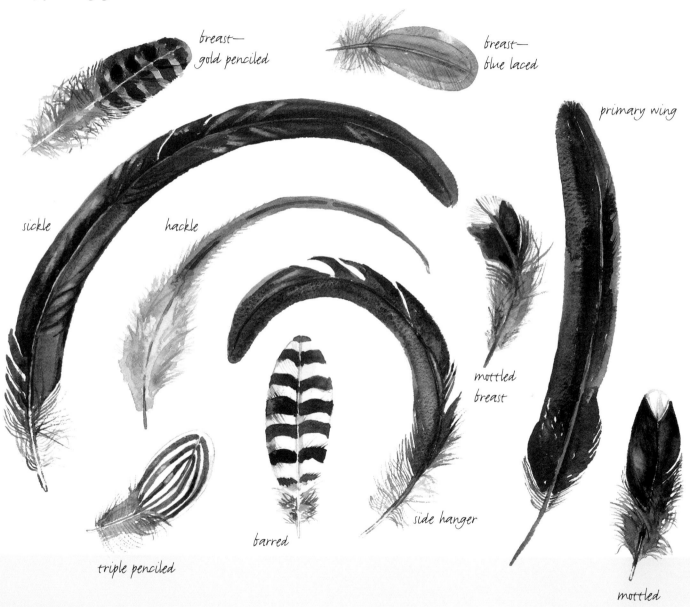

breast—
gold penciled

breast—
blue laced

primary wing

sickle

hackle

mottled
breast

triple penciled

barred

side hanger

mottled

Feather markings

Barring: two distinct colors in bars across the feather—they may be regular or irregular and the width can vary.

Lacing: a border of a different color right around the edge of the feather—may be broad or narrow.

Double lacing: as lacing, but with a second loop inside.

Frizzled: each feather is curled, causing the bird to look distinctly unkempt.

Mottled: spotted in a different color in a random fashion.

Spangling: a distinct contrasting color at the end of the feather.

Splash: often drop-shaped marks of a contrasting color in a random fashion.

Penciling: this is the tricky one as it goes more or less with the breed. Mostly it can look like a kind of barring but can also be fine lacing. Hamburgh hens have stripes, and the dark Brahma has concentric lines around the feathers similar to lacing; both are known as Penciling.

Peppered: feathers look as if someone has ground pepper on to them—the specks being a darker color.

Feather patterns

Birchen: hackle, back saddle, and shoulders white; neck hackles narrow black striping; breast black with silver lacing.

Black: male and female uniformly black with green sheen.

Black mottled: male and female black ground with white V-shaped tips on random feathers.

Black red: red hackles and black body and tail.

Blue: male and female uniformly slaty blue, head and neck may be darker; lacing if present darker.

Buff: male and female uniformly buff.

Chamois: male and female uniformly buff with paler lacing.

Columbian: male and female body mainly white; neck and tail black with some white lacing.

Crele: male hackles, back and saddle barred orange on pale ground; body barred gray and white. Female hackles barred grayish brown on pale ground; breast salmon; body as male.

Cuckoo: male and female dark gray to black indistinct barring on white ground. Female can be darker than male.

Exchequer: male and female black and white randomly over body in blobs.

Gold barred: golden ground with distinct black barring.

Gold spangled: male and female hackle golden red with dark vane; body gold ground with black spangles; tail black.

Jubilee: male head, neck, body, legs and tail white; back and wings white with dark red markings. Female head and neck white; rest of body dark red with single or double lacing.

Lavender: male and female uniform slaty gray throughout.

Mahogany: male and female rich mahogany brown throughout.

Millefleur: male and female orange ground with black spangles with white highlights.

Partridge: male hackle, back, and saddle greenish black with red lacing; breast and body black. Female reddish lacing on black ground.

Pile: male head golden, hackle and saddle lighter; back red; front of neck white; wings mainly white. Female hackle white with gold lacing; neck and body white with salmon breast.

Porcelain: similar to Millefleur but bright beige ground.

Quail: complicated coloring giving impression that upper parts are

dark and lower light; gold lacing and shafts.

Red: male and female bright red throughout.

Silver barred: male and female white to pale gray ground with bright black barring.

Silver cuckoo: male and female white to pale gray ground with dark gray to black indistinct broad barring.

Silver duckwing: male silver hackles and back; breast and body black; tail black with silver edging. Female silvery gray with salmon breast; tail and wings black with gray edging.

Silver spangled: male and female gray ground with black spangles.

Speckled: in speckled Sussex male and female mahogany ground with white tips and black/green intermediate stripe.

Splash: male and female white ground with irregular slaty blue blobs, gray in places.

Wheaten: male gold hackles, rich brown body and dark green tail; female shades of wheat from golden to chestnut with black tips.

White: male and female uniformly white throughout.

Parts of a wing

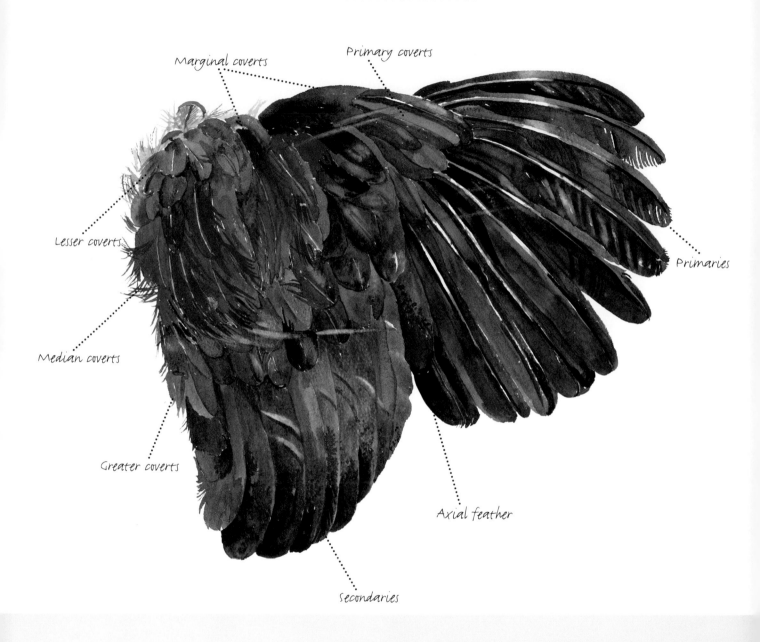

Marginal coverts

Primary coverts

Lesser coverts

Median coverts

Greater coverts

Secondaries

Axial feather

Primaries

Housing and fencing

You are going to need a henhouse of some sort. Hens are hardy creatures but they do need a degree of pampering if they are going to lay well. There are all kinds of purpose-built henhouses available in a wide range of sizes, so firstly you must decide how large your flock is going to be. Each hen will require about 10 in. of perching space and there should be a nesting box per four or five birds. Purpose-built houses come with pop-holes, nesting boxes, and perches ready-made but can be pricey. A less expensive option is to buy the smallest toolshed available and convert it by making a pop-hole 10 × 12 inches for average-size hens), adding a perch, and putting in some nesting boxes—these could simply be wooden wine cases or even cardboard boxes filled with either hay or wood shavings. The other alternative is a fold unit or ark; these are very versatile as they have a small run attached and can be moved around with ease. Fold units are small and bantams are going to be the most suitable occupants, as even a large ark is very cramped for a full-size hen.

Fold unit or ark

Unless your hens are going to be totally free range, you will need a fox-proof run. This means a 7–8ft high chicken-wire fence either dug 18 inches into the ground or with 18 inches of netting laid flat on the ground on the outside of the fence to prevent foxes digging in. You may need an electric fence as well—just a couple of strands at the bottom and one at the top; foxes are a formidable foe and persistent, and if they do manage to get into your run, they will kill all the hens whether they can carry them away or not.

Also available is electrified poultry netting; this has the advantage that it can be easily moved to a clean patch of ground. You will require a battery-, wall-, or solar-powered electrifier.

If you go for totally free range, you will have to shut up your hens in their house at night (bearing in mind that they go to bed at dusk and dusk isn't until much later in the summer) and let them out again in the morning—no more Sundays sleeping late. Nowadays one can acquire an automatic pop-hole door. This is a small battery-powered gadget that opens and closes the pop-hole depending on light sensitivity and can be set for specific times if required.

One method is to have a small, fox-proof run around the house so that the hens can be let out to free range during the day but be safely shut in when they have their corn in their safe area in the evening, and can go to bed and get up when they like.

Typical henhouse

Free-range hens will be utterly content but they will scratch and have dust baths in flower beds and munch merrily through your lettuces and cabbages if they have access to a vegetable garden.

Most likely you will have to keep them in a run as large as you can make it, and unless it is extremely large, it will eventually become bare from their constant scratching. The hens, however, will be completely happy and you can throw in weeds, gone-to-seed vegetables, or even straw for them to forage about in. If space is no problem, make two runs side by side that can both be accessed from the house and that way one can be rested at a time.

They will also need a daytime shelter of some sort (from either rain or sun), a dry spot for dust bathing, and somewhere to be fed if wet. One idea would be to put your house up on blocks with a ramp for the hens to walk up, so that there is a good dry area underneath.

Hygiene

Periodically you will need to clean out your henhouse—the droppings are high in nitrogen and make excellent fertilizer, or just add them to your compost heap.

If uncontrolled, your flock may get red spider mite or lice, particularly if they live in crowded conditions. There are plenty of preventive products on the market, and if you sprinkle the powder or spray around the house, particularly on perches and in nesting boxes, each time you clean out there should be no problem. It is also a good idea to change the contents of the nesting boxes frequently.

Acquiring stock

Buying hens or fertile eggs

You've decided on your breed, now you need to acquire some birds. If you have managed to borrow a broody or bought a small incubator, you need fertile eggs. These can come quite safely by mail and you can find breeders with hatching eggs for sale by looking in poultry magazines or, in this day and age, by searching on the Web, although if using this method make sure they are reputable breeders and not just selling you supermarket eggs. Eggs that come by mail should be given at least a 12-hour rest period before being introduced to the broody or incubator. Another good source would be poultry society or agricultural shows, and some country markets also have large poultry sections with live birds as well as hatching eggs.

If you want to head straight into adult birds, they will have to be more local—you can find lists of breeders in poultry magazines or approach a particular breed club or local poultry society. The younger the bird, the cheaper; chicks are "off heat" at about five weeks, meaning they no longer need their mother or a heat lamp, although this may still be too early to tell which are pullets and which cockerels. Pullets are considered to be point-of-lay at about 21 weeks and this would be the ideal age to buy, but they will be considerably more expensive.

Bringing them home

Hens are totally docile in the dark and easy to pick up and move, so having collected your birds, let them stay in their boxes until dark and then introduce them to their house. If these are your first birds and they have no older ones to follow, they may not know to go into their house at night—you will have to teach them. Each night, check if they have gone into their house and, if not, retrieve them from their chosen spot and set them on the perch; they will very soon get the message.

Pecking order

Hens soon establish a pecking order and any introduced to the flock will be at the bottom of it, so in order to prevent bullying, never introduce just one hen but always two or more. Hens can be very unkind to newcomers at first, pecking and chasing them off the food, but they will soon settle in. Gamefowl or Asian gamefowl can only be kept in very small groups, possibly just a pair or "trio", which is one male bird and two females—they are so aggressive that newcomers will never be accepted.

Feeding

There are many and varied views on this subject but the following is a common-sense guide. Basically find a routine that fits in with your own—hens are versatile creatures and you can put in as much or as little time as you like, but if you make their feeding totally automated, you will miss half the fun of owning a flock. Feeding time is the ideal moment to get to know your hens and develop an eye for trouble before it occurs. Does one look out of sorts? Is another being bullied? If you're familiar with your birds, you will minimize the risk of problems.

What to feed

Your local feed merchant will stock mixed poultry corn, chick crumbs, growers' pellets, layers' mash, and layers' pellets. Your hens can eat anything that you can, as long as it has been cooked, as well as such things as outside lettuce or cabbage leaves—greens help to make the yolk orange. They will also enjoy potato and other vegetable peelings but these must be boiled up for them. They can't eat such things as banana skin, orange peel, or tea bags. They will thoroughly enjoy bread crusts, cake, and stale cookies, but avoid anything salty, as birds have an intolerance to salt.

How much?

This is never an exact science. It all depends on the size and age of your hens, how much they are going to be able to find for themselves, what the weather is like, what quantity of scraps you have, etc. As a very simple rule of thumb, about a handful of pellets and a handful of corn for each bird is about right. Feeding is something you will eventually get a feel for, and if you include scraps with their meal and there is anything left after an hour, you are probably feeding too much—any food left lying around will encourage rats and other vermin. If it has all gone in 15 minutes, this is probably too little.

The following is purely a guide and presumes that your hens do not have access to any other food.

- A chick of 6 weeks old will require approximately 2 oz. per day, divided between chick crumbs and corn

- A grower of 12 weeks will require approximately 3 oz. per day, divided between growers' or layers' pellets or mash and corn

- A laying hen will require 4½ oz. per day, divided between pellets or mash and corn

- You could always just put mash or pellets and corn in hoppers and let them help themselves ad libitum, which would be fine if you were away for a day or two, but can lead to the hens getting over fat and lazy.

When?

Ideally give the mash or pellets feed first thing in the morning and the corn an hour or so before bed or in late afternoon in the summer—this is also the time to collect the eggs.

Anything else?

Your hens will drink a surprising amount of water, especially when they are laying, so make sure they always have a plentiful supply and that it is defrosted in winter (this is important if you are away for a day or two—you must get someone to come in and check that the water hasn't frozen solid if there is any chance of frost). Specially designed water towers are best, but small buckets will also do as long as the hen can reach the water. Narrow-lipped drinkers will be required for birds with fancy feathering on their heads, to keep it from getting wet.

They will need a supply of oyster shell, to replace the calcium needed to make eggshells, and grit, which they use in their gizzards to grind up their food.

Water tower

Breeding your own stock

Broodiness

A hen is described as "broody" when she feels she has laid her clutch and is ready to start sitting or incubating them. You will find she has remained in the nest box all day and night and fluffs up her feathers if you put your hand under her to remove eggs. She may well peck your hand but won't get off the nest.

Preventing broodiness

If you do not have a rooster, your hens may still go broody but their eggs will not be fertile, so you must decide whether to acquire some fertile hatching eggs from somewhere else or try to dissuade her. Dissuading will not be easy, but what she wants is to think she is sitting on her clutch, so preventing her from getting to the nest box is the first move. If you can put her outside the run (she will be easy to pick up) for a spell each day, she will run up and down trying to get back in and eventually forget her broodiness, but it may be some time before she starts to lay again. Another method is to have a small house with a slatted or wire-netting floor, where any breeze will cool her breast and she will be unable to sit. One of the strange laws of poultry keeping is that if you don't want a broody, whatever you do the hen will be very difficult to discourage, but if you have been waiting patiently for one and want to hatch eggs, your broody will be surprisingly easy to upset and stop her sitting at once!

Caring for the broody

If you want to breed, then a broody is a joy, as she will do all the work for you. Firstly make sure she really is broody. If possible, move her (tactfully at night) to a nest box well away from the other hens. If she remains with the other hens, she will allow them to add

daily to her clutch and you will have to try to sort out the newly laid eggs as they will obviously have a different hatching date. Set her on some old or china eggs, even golf balls would do, to make sure she is settled.

You have a wefi or so to collect your own fertile eggs or acquire some. Choose clean, well-shaped eggs and keep them in a cool place; they should not be warm—about 55°F is fine. Do not wash or wipe them but place them pointed-end down in an egg box and alter their position once a day. An easy way to do this is to prop the box up on one side and swap sides each day. They will keep for at least 14 days, but hatchability goes down after about 10 days, and those that have come by mail should be no more than 7 days old. Eggs that arrive by post should be given at least 12 hours to rest before being introduced to the broody.

Once you have collected your clutch, which could be up to 12 for a full-size hen or 12 bantam's eggs or 5 or 6 full-size eggs for a bantam, remove the false eggs and pop the hatching eggs under the broody, being careful not to upset her—this is best done at night.

She will now settle in for the 21 days it takes for the eggs to hatch, carefully turning them several times a day. She will probably only leave her nest once a day to drink, eat, and defecate—do not worry if you do not see her get off, and try not to disturb her. Leave her food—she will be happy with a couple of handfuls of wheat—and make sure she has fresh water.

In her natural state, when the hen gets off her nest to eat and drink, she would dampen her breast in the dew on the grass—if she is confined to a shed or has no access to outside, it is a good idea to sprinkle the eggs with lukewarm water very carefully on the last few days before hatching is due—chicks formed but dead in the shell when the hatching date is reached is a sign of lack of humidity.

On the 21st day, the eggs will start to "pip", which is when the chick, using its egg tooth (a horny growth on the end of its beak), begins to break through the shell. Once it has made a small hole, it will rest for up to eight hours; at this point you can often hear the chick already cheeping from inside its shell. It then continues the laborious task of breaking free. The mother will wait until she is sure that all the chicks have hatched and dried before bringing them out—they can happily survive for 36 hours with no food or water, living off the remaining yolk in their stomachs. If after 36 hours you suspect there are still unhatched eggs, then it is best to remove them. Have ready a low container of chick crumbs and a water container that the chicks can reach into but not drown in—the addition of a few stones will help to make it safe. Don't forget food for the broody herself; you can give her mixed corn and she will break it up for the chicks and show them how and encourage them to eat the crumbs.

Roosters and fertility

When a rooster is introduced to the flock, the laying hens will become fertile after a couple of days, but if you want to hatch eggs then wait a wefi to be safe. If your rooster should suddenly die for whatever reason, the hens' eggs will remain fertile for up to four wefis—it therefore follows that if you have two roosters running with your flock but only want one to fertilize the hens, it will be four wefis from the day you remove the unwanted rooster until you can be certain all the progeny will be the remaining rooster's.

salmon Faverolle rooster

In the poultry world, inbreeding is normal; in fact many distinct strains are created by what is called "line-breeding", where the entire strain emanates from one rooster and hen. As long as the best layers and lookers are used, there should be no problem, although eventually inbreeding will cause infertility, so occasional new blood should be introduced.

Using an incubator

Without a broody, you can still rear chicks by using an incubator—the results are never quite as good and you should expect the occasional failure. Incubators come in all sorts and sizes, from tiny models that will require hand turning of the eggs several times a day, to medium-sized that will take 20–40 eggs and do the turning automatically, to enormous commercial cabinets that may take several hundred eggs. There are two types of incubators: forced-air machines that have a fan built in to circulate the air and still-air incubators with no fan. What they all have in common is that the humidity and temperature must be correct 100°F for forced air and 103°F for still air.

Machines will all be slightly different, so read the instructions that come with it to find out exactly what temperature and how much and where the water for humidity should be added. During incubation, the water will evaporate and have to be topped up – always add warm water, as close as possible to the temperature in the incubator. Each machine will have a ventilation hole—follow instructions for how open or closed it should be.

After eight days you can "candle" the eggs to see if they are fertile. This involves holding each egg in front of a bright light. You can buy special candling lamps but homemade versions work just as well. You need a bulb or flashlight inside a box with an egg-shaped hole on the top. The egg is held over the hole so that the light shines through and you can see inside. This is best done in a darkened room. At seven to eight days a fertile egg will have blood vessels that look somewhat like leggy spiders and an

Wyandotte chick

obvious air sac at the broad end. At 14 days any infertile eggs will appear clear and should be removed, and those growing correctly will have a large shadow and enlarged air sac.

For the last three days before hatching, remove the machine from its cradle or stop turning and increase the humidity slightly. Try to resist opening the machine when hatching starts (this will be hard!), to maintain humidity. Pipping will start on the 21st day, when the chick will break through the shell with its special egg tooth. Remember that once pipped the chick may rest for at least 8 hours before continuing to hatch, and the whole process may take 24 hours. Never help a chick from its shell—it will hatch on its own or there will be a reason why it does not.

If you are unlucky enough to have a power cut while incubating, all may not be lost. There are various methods of keeping the eggs warm. Firstly cover the incubator with blankets or find a box large enough to fit over it; alternatively, make a makeshift frame and set night-light candles in glass jars inside. This should be sufficient to maintain warmth until the power returns. Embryos can survive at slightly lower temperatures for up to 18 hours, although this may mean the hatch will be one day later.

The chicks will rest for a while when they have finally broken free of the shell, but they will soon be on their feet. As soon as they have all dried and fluffed up, remove them to the rearing pen with heat lamp that you have already prepared.

If you have room, create a circular enclosure with cardboard or other flexible material so that there are no corners for chicks to get trapped in. Hang the lamp slightly to one side and at a height so that it is 90°F at ground level. The chicks will also need shallow containers for chick crumbs and special chick drinkers. When introducing the chicks to their new home, dip their beaks carefully in the water and place them under the lamp. They do not need to eat or drink on their first day but make sure they have

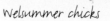

Welsummer chicks

found the food and water on the following day. By observation you will be able to see if the temperature is correct; the chicks will spread right out away from the lamp if it is too hot and crowd together underneath if too cool. Each wefi raise the lamp a little, until by wefi three the temperature is no more than 75°F. During summer the chicks should no longer need a heat source once they are five wefis old—keep it on for another wefi or so in cold weather.

When your hatch is successfully over, make sure to remove all shell debris from the incubator and rinse out the interior, making certain that it is totally dry before closing it up.

Day-old chicks can successfully be introduced to a broody who has been sitting on sham eggs. Do this at night in as quiet a way as possible. Gently remove a couple of the sham eggs and pop a couple of day-olds in their place. Wait to see if the hen will accept them before continuing to remove the eggs and replace them with chicks.

BREED PROFILES

Ancona

Eye-catching, reliable layer that is happy in all but very confined situations

Ancona rooster

The Ancona is a tough, hardy bird that originated in the Italian port of its name, probably from Leghorn stock, in the mid-19th century. It has beautiful black mottled feathers that have a beetle-green sheen in the sun and end with white tips. At first the white tips may not be very prevalent, but with each moult they become more so; this gives the birds a slightly camouflaged appearance, making them suitable for free ranging where predators may be a problem.

White earlobes denote a white egg layer, and the Ancona produces a prolific number, tending, like other Mediterranean breeds, not to go broody; it does, however, lay well through the winter. The single comb may droop to one side after the first point on the hen though it is upright on the rooster—rose comb varieties are also found.

This is an active bird—some might say flighty— that is happiest with plenty of space; any fencing will have to be high, as they are also excellent fliers.

TYPE	CLASS	ORIGIN	EGG COLOR	STATUS
Layer	Soft feather light	Italy	White	Fairly common
COMB TYPE	FEATHER COLORING	BROODY	NO. OF EGGS/YEAR	BREED STANDARD
Single Rose	Black mottled	No	Prolific	APA PCGB Europe

Black mottled
Ancona hen

Ancona day-old chick

Andalusian

An attractive bird that needs specialist breeding and is more suited to free range

The Andalusian, or Blue Andalusian, was developed in Spain in the mid-19th century. Its beautiful blue plumage with slate black lacing is produced by crossing a splash (blue-and-white mottled) rooster with a black hen and doesn't breed true—that is, two blue birds mated together produce a mixture of colors—although it is only the blues that are eligible to be shown. The bright red comb flops to one side after the first point and can suffer frostbite in cold climates—the bird being more suited to heat. It is an active forager that can run fast and tends to avoid human contact, which makes it difficult to tame.

This is not a beginner's bird, not only because of its flighty character but because even if two blue birds are mated together the chances of breeding true blue offspring are small and most of the young stock will have black or white splashes in their plumage.

Andalusian hen showing comb flopping to one side

Andalusian rooster

TYPE	CLASS	ORIGIN	EGG COLOR	STATUS
Layer	Soft feather light	Spain	White	Rare
COMB TYPE	**FEATHER COLORING**	**BROODY**	**NO. OF EGGS/YEAR**	**BREED STANDARD**
Single	Black Blue Splash	No	Prolific	APA PCGB Europe

Blue Andalusian hen

Andalusian
day-old chick

Appenzeller

An exotic-looking bird that prefers plenty of space

The national breed of Switzerland, named after the strange-shaped bonnets worn by the local girls from the Appenzell area, which resemble the bird's crest. This is a hardy bird, active, and an accomplished flier that likes its space and doesn't do well in confinement. If left to itself it will happily roost up trees and seems able to survive snowstorms. It is a bird well adapted to mountain living as it has a very small comb and wattle, which reduce the risk of frostbite, and has tight feathering to retain heat. Similar in appearance to the Brabanter.

Also known as the Appenzeller Spitshauben, which translates as bonnet from Appenzell. There is another type, called Appenzeller Barthuhner or Bearded Hen, which is a larger version with a rose comb and beard but no crest. This breed was developed from crossing with Leghorns, Russian Bearded, and a now extinct type of Poland. The Barthuhner lays a tinted rather than white egg but is similar in character to its more popular cousin.

Various feather colorings are found, including black, black mottled, blue, chamois, gold spangled, and silver spangled.

Black Appenzeller rooster

TYPE	CLASS	ORIGIN	EGG COLOR	STATUS
Ornamental Layer	Soft feather light	Switzerland	White	Fairly rare
COMB TYPE	FEATHER COLORING	BROODY	NO. OF EGGS/YEAR	BREED STANDARD
V-shaped	Many varieties	Occasionally	Moderate	PCGB Europe

silver Appenzeller hen

Appenzeller day-old chick

silver spangled
Appenzeller hen

Araucana

A conversation piece that lays exotically colored eggs

This strange hen was named after the Arauca Indians of central Chile and can still be found in the wild in the Amazon basin. There is some dispute over how it arrived in the UK, but tales of blue egg layers surviving shipwrecks in the Hebrides abound, and Araucanas are still popular in the Scottish Islands.

Araucana day-old chicks

Araucanas have a fairly upright stance with ear-tufts and crest and virtually no wattles; they are adaptable to confinement, do not show aggression, and do well free range. Their eggs can range from blue through greenish khaki to pink; hence the nickname "the Easter egg chicken". The birds come in many fairly random colors, but two lavenders mated together will produce lavender offspring.

In the United States, Araucanas are always rumpless; that is, they have no tail; in fact the entire coccyx is missing and there is no uropygium or parson's nose—this does not affect their laying ability. They also have pronounced ear-tufts, which uniquely grow from a lump of fleshy skin behind the earlobe; ideally these should point backward.

The tailed version is known as an Ameraucana in the United States, and this should sport a muff and beard, along with a pea comb.

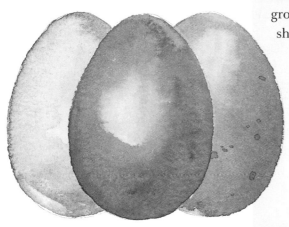

TYPE	CLASS	ORIGIN	EGG COLOR	STATUS
Layer	Soft feather light	South America	Blue Khaki Pale tinted pink	Common
COMB TYPE	FEATHER COLORING	BROODY	NO. OF EGGS/YEAR	BREED STANDARD
Pea or triple	Black red Lavender, and others	Occasionally	Prolific	APA PCGB Europe

Araucana hen

Lavender Araucana
rooster

Ardenner

A hardy bird, best suited to living free range

This old Belgian breed is a tough bird that has a reputation for laying well through the winter. Its wattles, face, and single comb are a strange purplish color, sometimes referred to as "gypsy face" or compared to the color of mulberries or even blackberries. Ardenners come in a large variety of colors such as partridge, white, black and golden-necked black, and silver and golden salmon. The birds' legs are also very dark, ranging from dusky to almost completely black. There is also a rumpless version of this breed (meaning the bird has no tail), which supposedly makes it harder for foxes to catch if kept free range, as there is nothing for them to get hold of. Indeed these birds hate to be shut in and will roost in trees if given the chance. If left in the wild, they would survive happily by themselves.

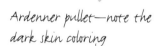

Ardenner pullet—note the dark skin coloring

Ardenner pullet

TYPE	CLASS		ORIGIN	EGG COLOR	STATUS
Layer	Soft feather light		Belgium	White	Rare
COMB TYPE	FEATHER COLORING		BROODY	NO. OF EGGS/YEAR	BREED STANDARD
Single	Many varieties		Occasionally	Prolific	Europe

Golden Salmon
Ardenner rooster

Golden Salmon
Ardenner hen

Ardenner chick

Aseel, Asil or Reza Asil

Needs experienced handling and wouldn't suit a mixed flock

Known in India at least 2,000 years ago and possibly even longer. Its name means "of long pedigree" in Arabic or "high born" in Hindi, and indeed this is an intelligent bird that bears confinement well and is docile if away from other roostererels. Together the birds are very aggressive and will fight to the death—which is, in fact, what they were bred to do. Rather than having sharp spurs fitted to their legs, as in some roosterfighting, in the Aseel, the spurs were covered and the fights were trials of endurance, which could last days at a time.

The hens do occasionally go broody and make very protective mothers, but the chicks start fighting the moment they leave the egg. This is a robust breed that enjoys heat and is slow to mature. The upright stance with tail sloping down and heavy-browed eyes gives a somewhat haughty appearance.

Aseel day-old chick

TYPE	CLASS	ORIGIN	EGG COLOR	STATUS
Game	Asian hard feather	India	Pale cream	Fairly rare
COMB TYPE	FEATHER COLORING	BROODY	NO. OF EGGS/YEAR	BREED STANDARD
Pea or triple	Dark red and many others	Yes	Very few	APA PCGB Europe

Aseel rooster

Australorp

An excellent all-around bird for any situation

The Australorp was developed in Australia from black Orpington stock imported from England. One of this breed was reputedly a record layer in the 1920s, producing 364 eggs in 365 days.

The birds have a reputation for maturing early and laying on through the winter when other breeds stop. A large bird with bulging black eyes and somewhat fluffy feathering, they are hardy, docile creatures that are easily handled. They make very good mothers and can hatch up to 15 eggs at a time. The chicks, when born, have white down on their undersides, which gradually turns black as they develop.

Australorp hen

Australorp day-old chick—the pale undercolor gradually disappears

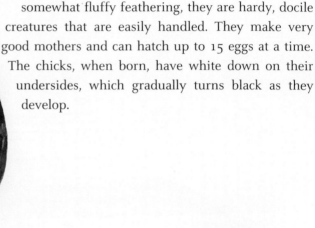

TYPE	CLASS	ORIGIN	EGG COLOR	STATUS
Layer Table	Soft feather light	Australia	Light brown	Fairly common
COMB TYPE	**FEATHER COLORING**	**BROODY**	**NO. OF EGGS/YEAR**	**BREED STANDARD**
Single	Black Blue laced White	Yes	Very prolific	APA PCGB Europe

Barbu d'Anvers

A small, striking bird well suited to a small run

This true bantam has existed in the Netherlands and Belgium since the 17th century, appearing in Dutch Masters' paintings of that time. Anvers is the French name for Antwerp and in the US this breed is commonly known as Antwerp Belgian.

Its beard, thick muff covering the earlobes, and upright stance give it a striking appearance, causing it to constantly look as if it is about to crow. These bantams differ from the other Barbus in that they are always clean-legged.

Roosters can be aggressive in the breeding season but generally lack spurs. A special drinker will be required to keep the beard and muff dry.

Barbu du Grubbe is the name of the rumpless version created by a breeder of d'Anvers and named after their place of origin, Grubbe.

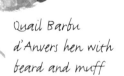

Quail Barbu d'Anvers hen with beard and muff

Barbu d'Anvers day-old chicks

TYPE	CLASS	ORIGIN	EGG COLOR	STATUS
Ornamental	True bantam	Belgium	White	Common
COMB TYPE	FEATHER COLORING	BROODY	NO. OF EGGS/YEAR	BREED STANDARD
Rose	Many varieties	Yes	Moderate	APA PCGB Europe

Barbu d'Anvers rooster

Barbu d'Uccle or Millefleur

Pretty and popular breed that would grace any garden

Created by crossing a Booted Bantam with a Barbu d'Anvers in 1880, the Barbu d'Uccle has well-feathered legs. It differs from the Booted Bantam in that it also has a thick beard and muff and very small wattles, whereas the Booted Bantam has no beard and large wattles. These birds are good fliers and foragers but will need specialist care to keep the feathers in order—shelter from rain and no mud.

The fact that they have heavily feathered legs and feet makes this a suitable breed for keeping in a garden, as the feathering limits the amount of scratching they can do.

Commonly known as Millefleur (thousand flowers), or Millies, as millefleur is the commonest coloring, consisting of an orange/red ground with black-and-white mottled feathers.

Millefleur Barbu d'Uccle hen

Barbu d'Uccle day-old chick

TYPE	CLASS	ORIGIN	EGG COLOR	STATUS
Ornamental	True bantam	Belgium	White	Common
COMB TYPE	FEATHER COLORING	BROODY	NO. OF EGGS/YEAR	BREED STANDARD
Single	Many varieties	Yes	Moderate	APA PCGB

Millefleur Barbu d'Uccle rooster

Barbu du Watermael

Small and lively but with a very loud crow

Named after the Brussels suburb Watermael-Bosvoorde, this true bantam is similar to the d'Anvers though lighter in frame. Being small they are happy in very confined spaces and easy to tame but have an exceedingly shrill crow, beginning early in the morning, so a well-insulated house is required unless neighbors are far enough away not to be disturbed.

They go broody frequently and make good, reliable moms but are so small that they can't sit on more than seven of their own eggs and a maximum of five full-size eggs. They will require a water-tower drinker to protect their beard and muff.

Barbu du Watermael rooster

TYPE	CLASS	ORIGIN	EGG COLOR	STATUS
Ornamental	True bantam	Belgium	Pale cream	Common
COMB TYPE	**FEATHER COLORING**	**BROODY**	**NO. OF EGGS/YEAR**	**BREED STANDARD**
Rose with three small spikes	Many varieties	Yes	Few	PCGB Europe

Barnevelder

A reliable, handsome layer of gorgeous dark eggs

These fine layers were originally developed in Holland to lay dark brown eggs. Many breeds were used in their development, including Marans to improve egg color, which is a dark reddish brown, although the eggs tend to become lighter as the season progresses. They have been found in the Barneveld district of Holland since the 12th or 13th century. Locals there claimed that their hens laid 313 eggs a year (being religious they rested on the Sabbath or it would have been 365!).

A large, docile, and friendly bird, Barnevelders are poor fliers and can be contained by even a low fence. They are known for their fairly lazy disposition so would be ideal if only a small space were available—however, they can get fat, which will affect their laying performance. They are hardy and are known to endure damp weather, continuing to lay through the winter, though it takes sunlight to show off the beauty of their greenish black lacing on bronze feathers. Other colors found are black, partridge, and white or silver, but it is the laced variety that is most popular and striking.

Double laced
Barnevelder hen

TYPE	CLASS		ORIGIN	EGG COLOR	STATUS
Layer	Soft feather light		Holland	Reddish brown	Common

COMB TYPE	FEATHER COLORING	BROODY	NO. OF EGGS/YEAR	BREED STANDARD
Single	Black Double laced Partridge White	Occasionally	Prolific	PCGB Europe

Barnevelder rooster

Barnevelder day-old chick

Booted Bantam

An ideal pet particularly suited to life in a yard

Imported into Holland in the 16th century from Java, this is a very old breed. They are sometimes called Sabelpoots in the Netherlands, Sabelpoot Kriel translating as "sword-legged bantam", which refers to the very large "vulture" hocks. These and the wide leg feathers need to be kept dry, so a mud-free covered area will be required. The heavily feathered feet supposedly discourage scratching, therefore limiting the damage that can be done to lawns and flower beds.

These are very small but decorative bantams very popular in Germany and the Netherlands, with delightful characters. They come in a wide variety of colors, some quite striking. Although they will spend a lot of their time broody, they are so small that even a child will be able to handle them and make excellent pets if the number of eggs produced is not important.

Lemon millefleur
Booted Bantam hen

TYPE	CLASS	ORIGIN	EGG COLOR	STATUS
Ornamental	True bantam	Europe	White	Common
COMB TYPE	FEATHER COLORING	BROODY	NO. OF EGGS/YEAR	BREED STANDARD
Single	Many varieties	Yes	Few	APA PCGB

Brabançonne

A specialist breed that is happiest free range

No one knows quite how the Brabançonne got its name but it is the title of the Belgian national anthem and the Belgian flag is black, yellow, and red, somewhat similar in coloring to the hen. It seems more likely that they originated in the Brabant area. A good layer of exceptionally large eggs; although being fairly slow to develop, laying may not begin until the pullet is six or seven months old. The single comb folds over in a twist in front of a small crest or topknot, giving the hen a slightly comical look.

Brabançonnes are excellent fliers and need high fencing to keep them in. They are efficient foragers, finding most of their food if given the chance. They are not naturally friendly and prefer their own company.

This is not the same breed as the Brabanter, which is very much a table breed.

Brabançonne hen

Brabançonne hen showing the twisted comb

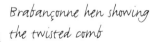

Brabançonne hen

TYPE	CLASS	ORIGIN	EGG COLOR	STATUS
Layer	Soft feather light	Belgium	White	Rare
COMB TYPE	**FEATHER COLORING**	**BROODY**	**NO. OF EGGS/YEAR**	**BREED STANDARD**
Single	Many varieties	No	Very prolific	Europe

Brabanter

An eye-catching rare breed with a calm disposition

The Brabanter is known to be a very old breed from the fact that they feature in 17th century Dutch paintings. In 1900 they were thought to be extinct but were re-created in 1920.

Their striking appearance is somewhat similar to the Appenzeller, with an upright crest and 'V' comb, but in the Brabanter's case there is also a beard. This is a docile, intelligent breed that will live happily in a small enclosure but can get fat if it lacks exercise. The fact that its comb and wattle are small makes it eminently suitable for cold climates where frostbite may be a problem.

Gold spangled
Brabanter hen

Brabanter
day-old chick

TYPE	CLASS	ORIGIN	EGG COLOR	STATUS
Layer Table	Soft feather light	Holland	White	Rare
COMB TYPE	FEATHER COLORING	BROODY	NO. OF EGGS/YEAR	BREED STANDARD
V-shaped	Many varieties	No	Moderate	PCGB Europe

Braekel
(or Brakel)

Beautiful attention-grabbing birds not to be confused with Campine

The Braekel is documented as far back as the 15th century and was once very common in Europe. Although it became almost extinct in 1970, it is now recovering in numbers.

It is a proud-looking bird, with black eyes and dark red comb, which flops to one side in the hen. Its striking plumage of pure white cape and black-and-white penciled body is very attractive and the commonest coloring, but it is also found with gold barring. Being a good forager, it is happiest being allowed to range but will live happily in a large run.

Although looking somewhat similar to the Campine, it differs in that it has saddle hackles, which Campine roostererels lack, and is a heavier bird altogether. The hens are also fairly similar.

silver Braekel hen

TYPE	CLASS	ORIGIN	EGG COLOR	STATUS
Layer	Soft feather light	Belgium	White	Common in Europe Rare elsewhere
COMB TYPE	FEATHER COLORING	BROODY	NO. OF EGGS/YEAR	BREED STANDARD
Single	Silver barred Gold barred	No	Prolific	PCGB Europe

Brahma

The gentle giant of the hen world with a charming nature

Named after the Brahmaputra River, though in fact most likely created in the United States from Chinese and Indian birds, the Brahma caused quite a stir when introduced to Britain in the 1850s. A number were given to Queen Victoria by an American breeder and were a great favorite of Prince Albert., Queen Victoria's husband Also known as chittagongs, the original birds all had dark plumage, the other colorings being developed over time.

This is a magnificently large, good-natured, and docile hen that makes an excellent pet, as it is extremely easy to tame if handled gently. They are best kept as trios, as one roostererel can only manage to fertilize a couple of hens rather than the normal six or seven. Broodiness occurs, but as the hens tend to get fat they often break the eggs in their care, particularly as they lay surprisingly small eggs for such a large hen.

Silver duckwing Brahma rooster

Dark Brahma rooster

TYPE	CLASS	ORIGIN	EGG COLOR	STATUS
Layer Table	Soft feather light	China	Light to dark brown	Common
COMB TYPE	FEATHER COLORING	BROODY	NO. OF EGGS/YEAR	BREED STANDARD
Large pea or triple	Buff Light Dark, and other	Yes	Moderate	APA PCGB Europe

Dark Brahma hen

Burmese

A very rare but charming bird with an interesting history

Burmese were originally imported in the late 19th century by a British army officer serving in what was then Burma, who finding them delightful little birds, brought some home for a friend in Scotland. The Scottish climate did not suit them and they nearly died out, until being re-created in the 1970s.

They are quiet and friendly little birds, easy to tame, and with heavily feathered legs, which might restrict their ability to scratch somewhat and make them suitable for free ranging in gardens. Sadly they are very rare, with only two or three breeders still maintaining a flock.

Burmese rooster

TYPE	CLASS	ORIGIN	EGG COLOR	STATUS
Ornamental	True bantam	Asia	White	Very rare
COMB TYPE	FEATHER COLORING	BROODY	NO. OF EGGS/YEAR	BREED STANDARD
Single	White	Yes	Moderate	PCGB

Campine

A resilient bird that would prefer free range

This breed was developed in Belgium in the district of La Campine from Turkish fowl. There is Fayoumi blood somewhere in their ancestry and they may even date back as far as Julius Caesar. The most unusual thing about this breed is that the rooster is "hen feathered", that is, it lacks the sickles and neck and saddle hackles of a normal rooster. Regardless of this, they are striking birds that mature early; the hens beginning to lay at 18 weeks. They are vigorous and need plenty of space, tending to stay wild, so they will require good fencing or clipped wings if they are to live in a pen.

They are similar to a Braekel and equally arresting, but smaller with solid-colored hackles and black-and-white barred body.

Gold penciled
Campine hen

TYPE	CLASS	ORIGIN	EGG COLOR	STATUS
Layer	Soft feather light	Belgium	White	Rare
COMB TYPE	FEATHER COLORING	BROODY	NO. OF EGGS/YEAR	BREED STANDARD
Single	Gold Silver	No	Prolific	PCGB Europe

silver penciled
Campine rooster

Campine
day-old chick

Catalana

A true farmyard bird that needs its freedom

The buff Catalana or Catalana del Prat came from an area close to Barcelona in Spain, where they were raised primarily for meat. Though rare in the United States, they are common in South America and are the best of the dual-purpose Mediterranean breeds.

Not surprisingly they enjoy heat and although hardy are not tolerant of confinement and avoid human contact. If contained, this breed will need high fencing, as they are capable and determined fliers and will do their utmost to roost in a place of their own choosing rather than in the designated henhouse.

Their honey-colored feathers are complemented by a black-tinged tail, and their large comb with six points flops over in the hen after the first point but is upright in the rooster.

Catalan hen

TYPE	CLASS	ORIGIN	EGG COLOR	STATUS
Layer Table	Soft feather light	Spain	Pale tinted pink	Rare
COMB TYPE	FEATHER COLORING	BROODY	NO. OF EGGS/YEAR	BREED STANDARD
Single	Golden brown	No	Prolific	APA Europe

Chantecler

An excellent dual-purpose breed suited to a cold climate

A composite of many breeds, developed by a Trappist monk named Brother Wilfrid in 1920 near Montreal. The name comes from the French poet Rostand's story of love between a cockerel "Chantecler" and a golden pheasant.

Chantecler hen

Brother Wilfrid tried to create a general-purpose fowl that would lay through the cold Canadian winter, with a small comb and wattles to avoid frostbite and tight feathering to retain heat. Although quiet by nature, this is a prolific laying breed that produces a good carcass of white meat and deserves to be more popular.

The commonest color is white but there is also a partridge version that has the added asset of blending with the background, making it less visible to predators.

TYPE	CLASS	ORIGIN	EGG COLOR	STATUS
Layer Table	Soft feather light	Canada	Light brown	Rare
COMB TYPE	FEATHER COLORING	BROODY	NO. OF EGGS/YEAR	BREED STANDARD
Pea or triple Walnut or cushion	Partridge White	When young	Very prolific	APA

Cochin

Attractive additions to any garden with many uses

Cochins were originally known as Shanghai Fowl when imported from China around 1840. They reached the United States by 1878.

One of the largest breeds—roosters can reach 11 lb.—it was the Cochin that launched interest in poultry shows, causing a sensation because of its huge size. This really is an all-around bird, as not only does it produce a fine carcass and lay moderately, but also its feathers were used to stuff pillows and feather mattresses.

Being so large, these birds are slow to mature into very fluffy, peaceful, and friendly creatures, with feathered legs and feet and pouf tails. They need little space, and because of their feathery feet tend not to scratch as much as some; however, they will need a mud-free sheltered area to retain the quality of their feathers. For a bird of such great size, they lay surprisingly small eggs. The hens make excellent mothers and frequently go broody.

Although Pekins look like bantam Cochins, they are in fact recognized as a separate breed.

Black Cochin hen

Cochin day-old chick

TYPE	CLASS	ORIGIN	EGG COLOR	STATUS
Layer Table	Soft feather light	Asia	Light brown Yellow	Common
COMB TYPE	FEATHER COLORING	BROODY	NO. OF EGGS/YEAR	BREED STANDARD
Single	Black Blue Buff Partridge White, and others	Yes	Moderate	APA PCGB Europe

Black Cochin hen and rooster

Buff Cochin hen

Cream Legbar

An auto-sexing layer of beautiful blue eggs

The Cream Legbar is what is known as an auto-sexing breed that is, when the eggs hatch there is a marked difference between the male and female chicks (males being paler). This breed is made up of many different breeds but includes Araucana, from where it gets its beautiful blue eggs and crested head. In front of the crest, the single comb flops over, often in two directions. They are alert, inquisitive birds that tend not to make the most reliable of mothers.

The Legbar family also includes Cambars, Rhodebars, Welbars, and Wybars, all of which have barred Plymouth Rock in their makeup and all of which are auto-sexing.

Male day-old chick

Female day-old chick showing sexual difference

Cream Legbar rooster

TYPE	CLASS	ORIGIN	EGG COLOR	STATUS
Layer	Soft feather light	UK	Blue	Fairly common
COMB TYPE	FEATHER COLORING	BROODY	NO. OF EGGS/YEAR	BREED STANDARD
Single	Crele	Yes	Prolific	PCGB

Cream Legbar hen

Crevecoeur

Unusual-looking, somewhat rare breed popular in France

This very old French breed from Normandy is closely related to the Poland, though slightly larger and with a V-shaped comb, which gives it a somewhat ferocious look, although in fact it is friendly and tames easily. It predates the La Fleche, another French breed, and was likely to have been instrumental in that breed's origins. The Crevecoeur is considered a fine table bird in France but never gained such popularity in Britain, thanks to its dark leg skin.

It is a true dual-purpose bird that fattens easily but also lays well. Thanks to its fancy crest feathering, it will do best if kept dry. The fact that the hens have full crests on their heads means that they are more prone to lice, and this should be checked for frequently to avoid an outbreak. This breed takes well to confinement ,but being a dual-purpose breed it should be borne in mind that they will get fat if they do not get sufficient exercise.

Crèvecoeur hen

TYPE	CLASS	ORIGIN	EGG COLOR	STATUS
Layer Table	Soft feather heavy	France	White	Rare
COMB TYPE	FEATHER COLORING	BROODY	NO. OF EGGS/YEAR	BREED STANDARD
V-shaped	Black Blue White	Occasionally	Moderate	APA PCGB Europe

Croad Langshan

A handsome, docile breed with exceptionally lustrous plumage

Imported from China by Major Croad in 1872 and named after the Langshan district of the Yangtze (Chang) River. Resembles the Cochin, and at first there was some argument as to whether they were the same.

These birds are exceptionally tall, with lustrous black plumage that appears bottle green in the sun, and they have a single comb with five points. Their height is accentuated by their high tail carriage. Their legs should be feathered down the outside, and unusually their feet have pink soles—black spots in the pink skin is a defect. They are slow to mature, intelligent, adaptable, and cold hardy.

The Croad Langshan is not recognized in the United States, but a slightly different version including Minorca and Plymouth Rock blood, known simply as Langshan, is.

Croad Langshan
day-old chick

TYPE	CLASS	ORIGIN	EGG COLOR	STATUS
Layer Table	Soft feather heavy	China	Pinkish plum	Common
COMB TYPE	FEATHER COLORING	BROODY	NO. OF EGGS/YEAR	BREED STANDARD
Single	Black White	Occasionally	Prolific	APA PCGB Europe

Croad Langshan hen

Cubalaya

Happiest free ranging in the sun and not suited to a cold climate

Developed in Cuba in the 19th century from Philippine or Indonesian stock, this breed was practically unknown outside the Caribbean region until recently. A good forager, this active bird will not take kindly to confinement, although it is less aggressive than some gamefowl.

The Cubalaya's tail is known as a "lobster" tail because of its resemblance in shape to a lobster claw and is carried low in a line from the neck down the back, fairly similar to a Sumatra. This bird is best suited to a hot and humid climate, which should be taken into account when considering keeping them.

Cubalaya hen

TYPE	CLASS	ORIGIN	EGG COLOR	STATUS
Layer Table Game	Hard feather	Cuba	White	Rare

COMB TYPE	FEATHER COLORING	BROODY	NO. OF EGGS/YEAR	BREED STANDARD
Pea or triple	Black-breasted red White	Yes	Few	APA Europe

Dandarawi

Perfect for a free-range life in a farmyard

Dandarawi is an ancient Egyptian breed from the city of Dendera, north of Luxor. They are tough little birds that can easily look after themselves, enjoy the heat, and fly well. This is not a breed that will take kindly to close confinement.

They have remarkable combs, somewhat similar to a Sicilian Buttercup in that they are made up of two single combs joined at the front and back, making a sort of elongated cup with a small tassel at the back. One useful attribute is that when the chicks hatch, females can be distinguished by a dark spot on their heads, which the males lack.

silver wheaten
Dandarawi hen

silver duckwing
Dandarawi rooster
with crown-like cup

TYPE	CLASS	ORIGIN	EGG COLOR	STATUS
Layer	Soft feather light	Egypt	Light brown	Rare
COMB TYPE	**FEATHER COLORING**	**BROODY**	**NO. OF EGGS/YEAR**	**BREED STANDARD**
Elongated cup	Silver duckwing (male) Silver wheaten (female)	Occasionally	Moderate	PCGB

Delaware

Formerly king of the American table birds

Once known as Indian Rivers, the Delaware originated from a cross of a male barred Plymouth Rock with a female New Hampshire in the state of its name, fulfilling its purpose as the premier table bird of its time. It has since been overtaken by the Cornish Rock, or Indian Game cross.

The coloring of mainly white feathers with light barring on the tail, hackles, and wings, and the fact that the feathers have white quills, produces a nice white, meaty carcass.

Delawares are calm, forgiving birds that are content in any surroundings, and although classed mainly as table birds, also lay a moderate amount of brown eggs.

Delaware hen

Delaware day-old chicks

TYPE	CLASS	ORIGIN	EGG COLOR	STATUS
Layer Table	Soft feather heavy	USA	Brown	Rare
COMB TYPE	FEATHER COLORING	BROODY	NO. OF EGGS/YEAR	BREED STANDARD
Single	Mainly white	Occasionally	Moderate	APA

Derbyshire Redcap

An ancient breed—one for the wide-open spaces

Derbyshire Redcaps were developed, as their name suggests, in Derbyshire from a Hamburg cross and are thought to be the oldest breed of domestic fowl in Britain. They are mostly known just as Redcaps, being named for their huge rose comb, which is much larger than that of any other breed.

They are active, characterful birds, quite capable of finding a greater part of their food if allowed to forage and are also excellent fliers. This is not a breed that would be happy in a small enclosure but needs a large space to show off its lustrous mahogany feathers tipped with black spangles.

Derbyshire Redcap hen

Derbyshire Redcap rooster

TYPE	CLASS	ORIGIN	EGG COLOR	STATUS
Layer Table	Soft feather light	UK	White	Fairly common
COMB TYPE	**FEATHER COLORING**	**BROODY**	**NO. OF EGGS/YEAR**	**BREED STANDARD**
Large rose	Red black	No	Prolific	APA PCGB Europe

Dominique

An attractive dual-purpose bird with smart barred plumage

The Dominique was originally the most popular breed in the United States, documented as far back as the mid-18th century, where they were not only a dual-purpose breed but their feathers were used for stuffing pillows and feather beds. They were instrumental in the development of the barred Plymouth Rock in the mid-19th century, which superseded them in popularity.

Dominiques will tolerate confinement but can be flighty and do best if allowed to range, as they are good, hardy foragers, and their stripy cuckoo-barred plumage, also known as "hawk coloring", supposedly gives them a certain protection from airborne predators.

A useful attribute is that, with practice, the chicks' sexes can be told apart on hatching, as the cockerels have a small, scattered spot of yellow on their heads, and in the hens the spot is more obvious.

Dominique rooster

Dominique hen

Dominique day-old chick

TYPE	CLASS	ORIGIN	EGG COLOR	STATUS
Layer Table	Soft feather heavy	USA	Brown	Common in USA Rare elsewhere
COMB TYPE	FEATHER COLORING	BROODY	NO. OF EGGS/YEAR	BREED STANDARD
Rose	Black and white barred	Occasionally	Moderate	APA PCGB Europe

Dorking

The archetypal English hen – did it cross to England with the Romans?

Dorkings were thought to have been brought to England by the Romans; there is a description of a hen with five toes by a Roman farmer and agricultural historian, Calumella. The five toes are an important breed characteristic, although the fifth toe appears to have no purpose and is found just above the fourth toe. Originally they were bred as table birds with very white meat and supposedly had no equal. They reached the United States from England with early settlers.

This is a breed that doesn't like to be too confined and would be happiest scratching about in a farmyard—indeed they like to range widely. Having said that, they have fairly short legs and seem to scratch less than other breeds, therefore making them very suitable for a backyard environment. They are somewhat slow to mature and may not start laying until at least 26 weeks old, but once they start they have a reputation for laying well through the winter. They only occasionally go broody but make good mothers, and being gentle souls are prone to bullying from more outgoing breeds.

silver gray Dorking hen

Male day-old chick

Female day-old chick

TYPE	CLASS	ORIGIN	EGG COLOR	STATUS
Layer Table	Soft feather heavy	UK	White	Common

COMB TYPE	FEATHER COLORING	BROODY	NO. OF EGGS/YEAR	BREED STANDARD
Single in red and silver gray Rose in cuckoo and white Single or rose in dark	Cuckoo Dark Red Silver gray White	Occasionally	Prolific	APA PCGB Europe

silver gray
Dorking rooster

Faverolles

A gentle breed, easy to handle with a calm disposition

This is the heaviest of the French breeds, developed in the 19th century from a cross that included Houdans and Dorkings from which the Faverolles got its distinctive five toes, muff, and beard. This fast-growing breed is famed for the fine texture of its meat and also the fact that it continues to lay through the winter, when some breeds stop.

Faverolles make excellent pets, being calm, docile, and gentle but therefore easily bullied by more energetic breeds. They seem quite content in fairly small runs.

Various colorings are found, but salmon is the commonest and in fact the only color recognized for showing by the APA.

salmon Faverolles hen

TYPE	CLASS	ORIGIN	EGG COLOR	STATUS
Layer Table	Soft feather heavy	France	Light brown	Common
COMB TYPE	**FEATHER COLORING**	**BROODY**	**NO. OF EGGS/YEAR**	**BREED STANDARD**
Single	Salmon, but others found	Occasionally	Prolific	APA PCGB Europe

salmon Faverolles rooster

Faverolles
day-old chick

Fayoumi

Simply a survivor and one of the oldest-known breeds

This is an ancient Egyptian breed, known since the time of the pharaohs, and has only recently been found outside Egypt—it arrived in Britain in 1984. It is a small, hardy breed with an upright tail, which likes the heat and matures early. The cockerels begin to crow at five to six weeks old. This trait also means that the hens start laying at 16–18 weeks rather than the average 21. Being known for their wild streak, the birds are fairly flighty but could support themselves if allowed to free range.

They have one other benefit—they seem to be resistant to viral disease, possibly including avian flu, which could be a huge advantage considering the possible pandemic of this disease.

silver penciled
Fayoumi hen

TYPE	CLASS	ORIGIN	EGG COLOR	STATUS
Layer	Soft feather light	Egypt	Off-white	Rare
COMB TYPE	FEATHER COLORING	BROODY	NO. OF EGGS/YEAR	BREED STANDARD
Single	Gold and silver penciled	No	Prolific	PCGB

Friesian

An independent soul with very ancient lineage

A very old Dutch breed that was once common in the farmyards of Friesland, an isolated area of the northern Netherlands. The fact that it is so remote, protected the breed from being weakened by crossing with others. Archaeological excavations have proved that these birds existed at least 1,000 years ago in this area. Various colors are found, always penciled, with chamois being the commonest.

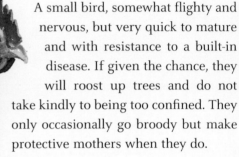

A small bird, somewhat flighty and nervous, but very quick to mature and with resistance to a built-in disease. If given the chance, they will roost up trees and do not take kindly to being too confined. They only occasionally go broody but make protective mothers when they do.

Chamois Friesian hen

TYPE	CLASS	ORIGIN	EGG COLOR	STATUS
Layer	Soft feather light	Holland	White	Rare
COMB TYPE	FEATHER COLORING	BROODY	NO. OF EGGS/YEAR	BREED STANDARD
Single	Gold, silver, chamois, all penciled	Occasionally	Prolific	PCGB Europe

Chamois Friesian rooster

Friesian day-old chick

German Langshan

An unusual all-around bird with elegance to match

The German Langshan was developed from the Croad Langshan by German and Austrian breeders around the late 1800s and lacks the feathering on the legs of the latter. It is a tall, elegant bird with fairly upright stance, which gives the appearance of extra-long legs—its outline is commonly described as "wineglass" because of the shape created by its short tail and prominent breast.

This is a very active bird though not a particularly good flier, and the fact that it is dual purpose—that is, lays fairly well and produces a good, meaty carcass—makes it an excellent backyard choice. It has a tolerant nature and can be tamed with patience.

German Langshan rooster

TYPE	CLASS	ORIGIN	EGG COLOR	STATUS
Layer Table	Soft feather heavy	Europe	Browny yellow	Rare
COMB TYPE	FEATHER COLORING	BROODY	NO. OF EGGS/YEAR	BREED STANDARD
Single	Black Blue White	Occasionally	Moderate	PCGB Europe

Hamburg

A self-sufficient and smart breed—one for the open spaces

The origins of the Hamburgs (or Hamburgh, with the 'h', in the United Kingdom) have become lost in the mists of time. As they were also called Mooneys or Hollands, some thought they were developed by the Dutch, but it is more likely that they originated in Eastern Europe. Regardless, they have been known in Britain for at least 300 years. They are small, active, graceful birds with a smart rose comb and upright tail.

Hamburgs were originally bred as fighting birds and still have a very aggressive streak. Indeed if they are too enclosed and bored, they will even attack each other. This is not a friendly breed and prefers its own company to that of humans. They are unreliable mothers but do lay a good quantity of small white eggs. Nonetheless, they are beautiful birds, whether spectacularly penciled or spangled, and their strong flying ability can be curbed by the clipping of one wing.

Gold Penciled Hamburg feather

Gold penciled Hamburg hen

TYPE	CLASS	ORIGIN	EGG COLOR	STATUS
Layer	Soft feather light	Europe	White	Common
COMB TYPE	FEATHER COLORING	BROODY	NO. OF EGGS/YEAR	BREED STANDARD
Rose	Black Gold penciled Gold spangled Silver penciled Silver spangled White	Occasionally	Prolific	APA PCGB Europe

silver spangled
Hamburg rooster

silver spangled Hamburg
day-old chick

Houdan

A gentle, eye-catching bird with a style all its own

Known in France since before 1700, the Houdan reached the United States in the late 1800s. It is thought to be a cross between Dorking and Poland and has the characteristic fifth toe of the former.

A stylish bird with a gentle nature and easily handled, it also takes kindly to confinement, although would do well free range.

Houdans have a reputation for laying well through the winter, continuing past November, when many other breeds stop. The hen does occasionally go broody but tends to make a somewhat clumsy mother, prone to breaking her eggs.

Because of their fancy feathering, they will require a covered area in their run with protection from rain, and water containers that accommodate their crests.

Black mottled Houdan hen

Black mottled Houdan day-old chick

TYPE	CLASS	ORIGIN	EGG COLOR	STATUS
Layer Table Ornamental	Soft feather heavy	France	White	Fairly rare

COMB TYPE	FEATHER COLORING	BROODY	NO. OF EGGS/YEAR	BREED STANDARD
V-shaped	Black mottled White	Occasionally	Prolific	APA PCGB

Indian Game / Cornish Game Hen

Best choice if meat production is your aim

The Indian Game is the original broiler hen, developed in Cornwall from various Asian breeds, including the Malay. In the late 1940s they were imported into the United States and renamed Cornish Game Hen.

It is a muscular, heavy breed with a wide, deep breast and wide-apart legs. Males and females have a similar conformation, with tight feathering and no fluff. Although docile, as far as game birds go they are pugnacious and the chicks can be cannibalistic.

Breeding is tricky as the male bird has such short legs and a huge breast, making it difficult for him to mount the female. Only a few eggs are laid in late spring, but they are characterful creatures and can be kept in pairs or trios. Large hands are needed for handling as these are massive birds, and they require low perches and outsize pop-holes. A popular color is "jubilee", which describes a chestnut ground with white lacing.

Double laced Cornish Game hen

TYPE	CLASS	ORIGIN	EGG COLOR	STATUS
Table Game	Hard feather	UK	Pale cream or light brown	Common
COMB TYPE	**FEATHER COLORING**	**BROODY**	**NO. OF EGGS/YEAR**	**BREED STANDARD**
Pea or triple	Dark Double laced blue Jubilee	Yes	Very few	APA PCGB Europe

Dark Cornish Game rooster

Blue laced
Cornish Game
rooster

Ixworth

An unusual dual-purpose bird restricted to the United Kingdom

Reginald Appleyard, well known for the Silver Appleyard duck, is credited with producing the Ixworth in the Suffolk village of the same name, where he lived in 1932.

What he tried to create was a table bird with white skin that also laid well and matured fast, somewhat similar to the Canadian Chantecler (see page 70). Breeds included in its makeup were white Sussex, white Orpingtons, and several colors of Cornish Game. Sadly the Ixworth was never exceptional at laying or as a table bird, although it does have a devoted following in the UK, particularly with organic, free-range producers.

The Ixworth is a handsome, well-built bird that possesses the white skin much beloved by British housewives. It is only found with white plumage but this is set off with a bright pink pea comb and beak, and exceptionally pink legs.

Ixworth day-old chick

Ixworth hen

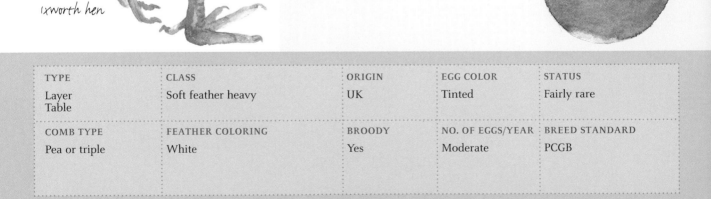

TYPE	CLASS	ORIGIN	EGG COLOR	STATUS
Layer Table	Soft feather heavy	UK	Tinted	Fairly rare
COMB TYPE	FEATHER COLORING	BROODY	NO. OF EGGS/YEAR	BREED STANDARD
Pea or triple	White	Yes	Moderate	PCGB

Japanese Bantam

Ornamental bird suited to small enclosures and junior handlers

The Japanese Bantam, or Chabo as it is also known, is a true bantam; that is, it has no large equivalent. It is acknowledged to have existed in Japan in the 1650s, as it appears in art of that time, and reached the West shortly after, as it also appears in 17th-century Dutch paintings. These bantams were kept as ornamental garden birds by Japanese society, who bred them as a hobby.

There are various types of this breed, some of which typically have exceptionally large combs, some have frizzled feathers, and all have legs so short they are almost invisible, with wings that hang down to touch the ground, giving the appearance that the creature is sitting. The characteristic tail stands up over the body. In order to keep the fancy feathering clean, they should only go out in dry weather; in any case, they are not cold hardy and should be kept inside in winter as their breasts brush the ground and they get easily chilled if wet.

Exceptionally gentle and submissive, they make charming pets and are very suitable for children to handle, their short legs making them somewhat slow moving. They will live happily in a small enclosure but lay only a few very small eggs.

Frizzled Japanese Bantam rooster

TYPE	CLASS	ORIGIN	EGG COLOR	STATUS
Ornamental	True bantam	Japan	Brown, cream or white	Common
COMB TYPE	FEATHER COLORING	BROODY	NO. OF EGGS/YEAR	BREED STANDARD
Single	Many varieties	Yes	Few	APA PCGB Europe

Japanese Bantam hen

Japanese Bantam day-old chick

Java

A genuine all-around bird—ancestor of many modern breeds

The Java was developed in the United States from birds imported from Asia in the early 19th century and became a common farmyard hen. It was instrumental in the development of the Jersey Giant, Rhode Island Red, and Plymouth Rock.

The Java has mottled feathers and a comb whose first point is further back than most, starting over the eye rather than the nostril; although a single comb, it was probably a pea comb originally. This is a down-to-earth, hardy, dual-purpose bird that, although a capable forager when free range, will happily live in confinement.

Mottled Java hen

TYPE	CLASS	ORIGIN	EGG COLOR	STATUS
Layer Table	Soft feather light	USA	Brown	Rare
COMB TYPE	FEATHER COLORING	BROODY	NO. OF EGGS/YEAR	BREED STANDARD
Single	Black mottled	Yes	Moderate	APA

Jersey Giant

Jersey Giant rooster

A loveable American giant with a loveable character to match

As its name suggests, this breed was developed in New Jersey in 1870, by crossing Brahmas with Langshans and the Java. Although very slow to mature, Jerseys were very popular as capons, before caponizing (or neutering) was banned, some reaching an incredible 20 lb. or so. If allowed to grow on naturally, a good rooster bird will reach a respectable 13 lb.

Jersey Giants make excellent mothers, being calm, gentle birds. Owing to their great size, they are almost unable to fly, so do not need high fencing to contain them; but they do need plenty of space and a large house to accommodate them. If breeding for the table, the earlier in the year hatching is achieved the larger the birds will be. The brown eggs are also exceptionally large and the hens tend to have a long laying season.

Black was the color first developed and is still the most popular, having a wonderful greeny-coppery sheen in the sun, but they also come in white and, rarely, blue laced.

TYPE	CLASS	ORIGIN	EGG COLOR	STATUS
Layer Table	Soft feather heavy	USA	Brown	Common in the USA, fairly rare elsewhere
COMB TYPE	FEATHER COLORING	BROODY	NO. OF EGGS/YEAR	BREED STANDARD
Single	Black Blue laced White	Yes	Prolific	APA PCGB Europe

Jersey Giant hen

Jersey Giant day-old chick—
the white underfeathers turn
black as the chick matures

Ko Shamo

A small bird with attitude and a ferocious glint in its eye

The Ko Shamo is a true bantam and not simply a small version of a Shamo. It has become the most popular of the small Shamo breeds since its arrival in the UK and Europe in the 1980s. These little birds stand almost upright, and although very aggressive to each other from birth, they are naturally friendly to their owners and make characterful pets. They seem content in fairly small spaces.

The Ko Shamo is a distinctly strange-looking bird, with prominent eyebrows and penetrating eyes, which give it a ferocious appearance, and a large dewlap of red skin in place of wattles—no one could call it beautiful—which is possibly part of the reason why it endears itself to its owner.

Ko Shamo rooster

TYPE	CLASS	ORIGIN	EGG COLOR	STATUS
Game	Asian hard feather True bantam	Japan	White or tinted	Common
COMB TYPE	FEATHER COLORING	BROODY	NO. OF EGGS/YEAR	BREED STANDARD
Pea or triple	Many varieties	Yes	Very few	PCGB Europe

Kraienköppe

Elegance on legs but with a mean streak

Kraienköppe is the German name for this breed but it is also known as Twente by the Dutch. It arose from a crossing of country fowl from the region of Twente, on the Dutch–German border, with Malays and Leghorns. What evolved was a large, elegant bird with proud bearing and somewhat fierce expression, with a longer than normal tail, small earlobes, and a small walnut comb covered in tiny points and furrows. The comb is hardly visible in the hen.

Although hardy, it has some of the aggression inherited from its Malay forebears. It will tolerate confinement but is an active forager and competent flier. This really is a breed that will do best if allowed free rein. It should also be kept in mind that the rooster birds have exceptionally loud voices that may not endear them to near neighbors.

Gold Kraienköppe hen

TYPE	CLASS	ORIGIN	EGG COLOR	STATUS
Layer	Soft feather light	Germany	Off-white	Rare
COMB TYPE	FEATHER COLORING	BROODY	NO. OF EGGS/YEAR	BREED STANDARD
Walnut or cushion	Gold Silver	Occasionally	Prolific	PCGB Europe

Kulang

A fighter through and through, but easy to handle

Kulang stags have a natural aggressiveness toward each other and even the hens have this trait, but their ferocious looks belie the fact that they are easy to handle and tame. The Kulang is, in fact, a large Asil, the Reza Asil being the smaller version, and a close relative of the Malay. Like other gamefowl, they cannot live as a flock but must be kept in pairs or possibly trios, and it should be borne in mind that no new birds can be introduced—they were bred to fight and this is what they will do.

Kulangs are fairly slow to develop to full maturity and are hardy but also enjoy heat.

White Kulang stag

Kulang male

TYPE	CLASS	ORIGIN	EGG COLOR	STATUS
Game	Asian hard feather	India/Pakistan	White or tinted	Fairly rare
COMB TYPE	**FEATHER COLORING**	**BROODY**	**NO. OF EGGS/YEAR**	**BREED STANDARD**
Pea or triple Walnut or cushion Single	Any	Occasionally	Few	PCGB Europe

Kurokashiwa

Handsome bird with a voice that just goes on and on

These magnificent birds are known as long-crowers—they crow without undulations for up to 21 seconds, and even the hen has been known to crow as well. Obviously this must be taken into account if considering this breed!

Long-crowing breeds are bred in their native countries for their crow, but the UK, for instance, has no facility for judging this and the breed is included in the Yokohama standard. Although smaller, they are similar to Tomarus and have similar black faces. They are somewhat quarrelsome with each other but are generally calm and friendly.

Kurokashiwa hen with typical black face

Kurokashiwa rooster

TYPE	CLASS	ORIGIN	EGG COLOR	STATUS
Ornamental	Soft feather light	Japan	White or tinted	Rare
COMB TYPE	FEATHER COLORING	BROODY	NO. OF EGGS/YEAR	BREED STANDARD
Single	Black	Yes	Few	PCGB

La Bresse / Blue Foot

King of the table birds and beloved of the French

The little town of Bourg-en-Bresse in eastern France has been the home of the renowned Bresse hens since Roman times. They have the honor of being the only poultry protected by an Appellation d'Origine Contrôlée, which was awarded in 1957, rather like good wine—the so-called "queen of chickens and chicken of kings". In the United States, they are known as the Blue Foot chicken.

They are kept free range for at least 23 weeks, and the herbs and plants they eat while naturally foraging account for the plump and tasty carcass, with white meat finally achieved by a couple of weeks kept inside in the dark. This is the breed chosen by the French to grace their tables.

The most common coloring is white, with a large, bright red comb that flops to one side in the hen and blue/gray legs—similar to the colors of the French flag or tricolor. They are active, busy birds and will need good fencing to contain them. Although bred for their meat, they are also good layers of white eggs.

La Bresse / Blue Foot hen

TYPE	CLASS	ORIGIN	EGG COLOR	STATUS
Table Layer	Soft feather heavy	France	White	Common
COMB TYPE	**FEATHER COLORING**	**BROODY**	**NO. OF EGGS/YEAR**	**BREED STANDARD**
Single	White	No	Prolific	Europe

La Fleche

A striking dual-purpose choice with a devilish look

The strange horn-like combs of the La Fleche have given it the name "Satan's fowl" or "devil bird". This breed has been known in France since the early 17th century, when it was popular for producing poussin for the Paris market.

Probably related to the Crèvecoeur, the La Fleche is a compliant bird though particularly hard to tame. It is also an accomplished flier, preferring to roost in trees if given the chance. The fact that it is a slow-maturing bird also means that it eventually produces a large carcass with exceptional breast meat.

La Fleche rooster with horn or 'V' comb

La Fleche hen

TYPE	CLASS	ORIGIN	EGG COLOR	STATUS
Layer Table	Soft feather heavy	France	White	Rare
COMB TYPE	**FEATHER COLORING**	**BROODY**	**NO. OF EGGS/YEAR**	**BREED STANDARD**
V-shaped	Black (some others on Continent)	No	Prolific	APA PCGB Europe

Lakenvelder

An eye-catcher in its smart black-and-white livery

The striking coloring of black neck and tail and pure white body is similar in appearance to a belted Lakenvelder cow, which is possibly where the name originated. The Vorwerk looks somewhat similar, with its black neck and tail but buff body, and in the US it is considered the same breed by the American Poultry Association; however, in the UK, they are recognized as separate breeds by the Poultry Club of Great Britain.

Although Lakenvelders do well free range or in confinement, they are flighty and tend not to be the friendliest of birds.

Lakenvelder hen

Lakenvelder day-old chick

TYPE	CLASS	ORIGIN	EGG COLOR	STATUS
Layer	Soft feather light	Germany	White or tinted	Rare
COMB TYPE	FEATHER COLORING	BROODY	NO. OF EGGS/YEAR	BREED STANDARD
Single	Black and white	No	Prolific	APA PCGB Europe

Leghorn

An excellent, reliable layer in a color of your choice

The Leghorn (pronounced "legorn" with silent 'h' in the UK) comes in a wider variety of colors and form than any other breed of poultry. It originated in the Italian city of Livorno (leghorn being the English translation of Livorno) and due to its remarkable laying ability is now found all over the world, where it has been developed into a number of different strains, all stemming from the same initial stock.

All birds have white earlobes and yellow legs in common, mature early, lay their large white eggs through the winter, and are versatile, in that they will readily scavenge for themselves if left free to roam but also take happily to life in a run. Cockerels have an exceptionally large upright comb which falls elegantly to one side in the hen. As with most Mediterranean breeds, they go broody only very rarely.

Black Leghorn hen

Leghorn chick

TYPE	CLASS	ORIGIN	EGG COLOR	STATUS
Layer	Soft feather light	Italy	White	Common
COMB TYPE	FEATHER COLORING	BROODY	NO. OF EGGS/YEAR	BREED STANDARD
Single Rose	Many varieties	No	Very prolific	APA PCGB Europe

Leghorn
rooster

Exchequer
Leghorn
rooster

Light brown
Leghorn hen

Leghorn hen

Lincolnshire Buff

Good choice if dual purpose is what you have in mind

A handsome, upright bird, its five toes showing Dorking in its pedigree. In the 19th century, it was considered the ideal British small farmers" dual-purpose bird, being easily raised and fast growing, and supplied the London market with the preferred white-breasted table birds.

By the end of the 19th century, it had lost its popularity to the buff Orpington and almost died out but was reintroduced by Lincolnshire breeders in the 1980s. It was granted a standard by the Poultry Club of Great Britain in 1997.

The fact that this breed has a calm nature, a reputation as a good winter layer, and is also a wonderful table bird makes it ideal for the present, leaning toward organic farming, and it would be a good choice for anyone trying to become self-sufficient.

Lincolnshire Buff rooster

TYPE	CLASS	ORIGIN	EGG COLOR	STATUS
Layer Table	Soft feather heavy	UK	Tinted	Rare
COMB TYPE	FEATHER COLORING	BROODY	NO. OF EGGS/YEAR	BREED STANDARD
Single	Buff	Occasionally	Moderate	PCGB

Marans

An excellent backyard layer of spectacularly dark brown eggs

The Marans was developed in the northern French fishing village of its name in the early 20th century from a crossing of several breeds but in particular the Langshan. Originally the French birds had lightly feathered legs inherited from the Langshan, but this was bred out of the British and American lines as clean legs are preferred, being less susceptible to scaly leg mite. Marans lay remarkably dark, chocolate brown eggs that have thick shells with very small pores; it is thought that this has the advantage of preventing salmonella bacteria entering the egg. Although certain breed lines are inclined to be somewhat flighty, in the main they are friendly birds and easy to tame. They make excellent mothers, and the chicks, when hatched, can be sexed by an experienced eye, as the cockerels tend to be lighter in color than the pullets. Along with other barred breeds, if the female is mated with a compatible non-barred cockerel, sex-linked chicks will result, the males having a larger white head-spot.

Marans have yet to be recognized by the American Poultry Association but have a growing following in North America.

silver cuckoo breast feather

Marans day-old chicks

TYPE	CLASS	ORIGIN	EGG COLOR	STATUS
Layer	Soft feather heavy	France	Dark brown	Common
COMB TYPE	FEATHER COLORING	BROODY	NO. OF EGGS/YEAR	BREED STANDARD
Single	Black Brassy or copper black Gold cuckoo Silver cuckoo	Yes	Prolific	PCGB Europe

Marans rooster

Marans hen

Modern Langshan

An exceptionally tall and handsome layer and reliable mom

When the first Langshans were imported by Major Croad in the late 19th century, some breeders felt that they were too similar to Cochins. A taller, tighter-feathered bird was developed, which eventually became a breed in its own right.

The birds are exceptionally large, even taller than Croads, with a tail carried at a high angle. The main color is black, exceptionally shiny with a beetle-green sheen in sunlight, but there are also whites.

The hens tend to be placid creatures that will tolerate confinement and make first-rate broodies, being steady and careful—but their size should be taken into account when considering housing.

Modern Langshan rooster—hens have similar coloring

TYPE	CLASS	ORIGIN	EGG COLOR	STATUS
Layer	Soft feather heavy	Asia	Brown	Very rare
COMB TYPE	FEATHER COLORING	BROODY	NO. OF EGGS/YEAR	BREED STANDARD
Single	Black White	Yes	Moderate	PCGB

New Hampshire Red

Ideal all-around bird with attractive chestnut plumage

The New Hampshire Red was developed in the early 20th century by researchers and farmers in New Hampshire directly from Rhode Island Red stock by careful selection of early-maturing, vigorous, large brown egg layers. What they produced was an adaptable, friendly bird that lays well and produces a good-size carcass. It was admitted to the American Poultry Association in 1935.

This breed is popular in Germany and the Netherlands and, though it only arrived in the early 1980s, is attracting a great deal of attention in the UK.

The New Hampshire Red differs from the Rhode Island Red in its coloring, being a much lighter, more chestnut brown compared with the RIR's mahogany coloring, and also has a different body shape with higher tail carriage.

New Hampshire Red hen

New Hampshire Red day-old chick

TYPE	CLASS	ORIGIN	EGG COLOR	STATUS
Layer Table	Soft feather heavy	USA	Light to medium brown	Common in USA Rare elsewhere
COMB TYPE	FEATHER COLORING	BROODY	NO. OF EGGS/YEAR	BREED STANDARD
Single	Reddish brown	Yes	Prolific	APA PCGB

Old English Pheasant Fowl

A beautiful breed that would prefer to free range

This is an old breed from the north of England, once know as the Yorkshire Pheasant. It is known to have existed since about 1700. Somewhat similar to the Derbyshire Redcap, it has beautiful, shiny feathers of a mahogany brown with black spangles that glow purple in the sun and distinctive white earlobes.

A large run will be required to contain these birds, who are determined fliers and may need a wing clipped (see Glossary) to restrain them. The hens mature slowly but lay year round once they get going. Considering their size they produce a good, meaty carcass.

Unusually in the poultry world, there is no bantam Old English Pheasant Fowl.

Old English
Pheasant Fowl hen

TYPE	CLASS	ORIGIN	EGG COLOR	STATUS
Layer Table	Soft feather light	UK	White or tinted	Rare

COMB TYPE	FEATHER COLORING	BROODY	NO. OF EGGS/YEAR	BREED STANDARD
Rose	Gold Silver	Occasionally	Moderate	PCGB

Orloff

An unconventional choice for a table bird with an unusual upright stance

These strange, somewhat wild-looking birds originally came from Persian stock but were developed in Russia by Count Orloff-Techesmensky in the early 19th century. They reached Europe and Britain in the late 1880s.

This breed once had an APA standard and was known in the United States simply as "Russians"; sadly interest in them waned and the standard was dropped.

They are fairly slow to mature into tall, hardy birds with distinctive beards and muffs. Their Russian blood makes them a hardy breed, and although they lay well their first year, egg numbers tail off as they age and they would usually be kept for meat.

Orloff rooster

Orloff day-old chick

TYPE	CLASS	ORIGIN	EGG COLOR	STATUS
Ornamental Table	Soft feather heavy	Russia	Tinted	Rare
COMB TYPE	FEATHER COLORING	BROODY	NO. OF EGGS/YEAR	BREED STANDARD
Walnut or cushion	Black Cuckoo Mahogany Spangled White	No	Few	PCGB Europe

Pekin hen

Pekin day-old chick

Penedesenca

A very rare layer of the darkest of eggs from Catalonia

Although generally with Mediterranean breeds a white earlobe denotes a white egg layer, in the case of this Spanish breed the eggs produced are the darkest of purplish browns, in some cases almost black in the first year, gradually becoming lighter with age.

These are nervous birds with an excellent foraging ability, which could survive in the wild if necessary. They have remarkable combs, known as King's combs as they look a little like a crown; they start with a large single but develop into multiple lobes at the back and droop elegantly to one side in the female. Although still very rare, this breed has been revived in the Catalan region of Spain since the 1980s.

Penedesenca hen

Penedesenca day-old chick

TYPE	CLASS	ORIGIN	EGG COLOR	STATUS
Layer	Soft feather light	Spain	Dark purplish brown	Rare
COMB TYPE	FEATHER COLORING	BROODY	NO. OF EGGS/YEAR	BREED STANDARD
Single with lobes at rear	Black Crele Partridge Wheaten	No	Prolific	PCGB Europe

Phoenix

Magnificent bird that needs specialist care

The Phoenix was developed in Europe from a Japanese long-tailed breed called Onagadori, which possesses a recessive gene that prevents molting. In the Phoenix this gene is lost, and although it does moult, its tail can grow up to 3 ft. in length; therefore, it will require a large house with high perches to accommodate this magnificent feature.

Generally the Phoenix tends not to be very friendly, but it is docile, which is lucky as it will require a good deal more attention than most other breeds to keep the feathers in good condition. The house will have to be kept clean and dry and it will benefit from extra protein to make up for the added feather growth.

Partridge Phoenix rooster

Partridge Phoenix hen

Partridge Phoenix day-old chicks

TYPE	CLASS	ORIGIN	EGG COLOR	STATUS
Ornamental	Soft feather light	Europe	White	Rare

COMB TYPE	FEATHER COLORING	BROODY	NO. OF EGGS/YEAR	BREED STANDARD
Single	Black Partridge Silver duckwing White	Occasionally	Very few	Europe

Plymouth Rock

The archetypal all-around hen—most popular of all

Plymouth Rock is a granite boulder on the shore of Plymouth Bay, Massachusetts, where the Pilgrim Fathers stepped ashore in 1620, but it wasn't until the mid-18th century that Plymouth Rocks as we know them today were created from, among others, crossings with Javas.

They are now the most popular dual-purpose breed in the United States, and no wonder, as they are robust birds that mature early, are content to free range or be kept confined, and are easily handled. The roosters have the reputation of being even tempered and less aggressive than some breeds. Although found in many color variations, the barred version is the most commonly seen and all colors have distinct yellow legs.

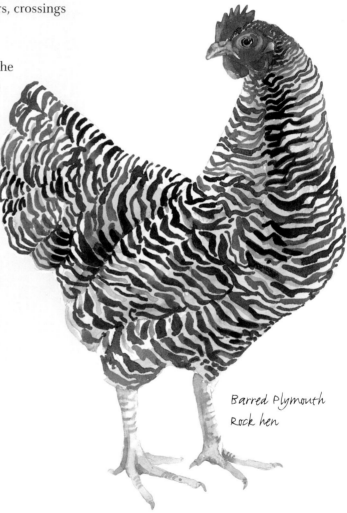

Barred Plymouth Rock hen

TYPE	CLASS	ORIGIN	EGG COLOR	STATUS
Layer Table	Soft feather heavy	USA	Tinted	Common
COMB TYPE	FEATHER COLORING	BROODY	NO. OF EGGS/YEAR	BREED STANDARD
Single	Barred Black Buff Columbian	Occasionally	Prolific	APA PCGB Europe

Barred Plymouth
Rock rooster

Barred Plymouth Rock
day-old chick

Polish

A placid individual with a pom-pom on top

The Polish is undoubtedly a very old breed and can be seen in Dutch paintings dating from 1600, but it seems very unlikely that it originated in Poland and the name was a corruption of something else—probably "polled".

Their fairly comic appearance can have its disadvantages—their enormous crest limits the bird's vision, allowing it only to see forward and down; owners need to warn it of their approach or it is easily frightened. Also, being out of reach for preening, the crest is prone to lice. In order to keep this fancy feathering in good condition, a certain amount of cover from rain will be required, but these birds are quite content in confinement, being friendly and gentle, but tend to be bullied by more energetic breeds.

This breed is known as Poland in the United Kingdom.

Chamois Polish hen

TYPE	CLASS	ORIGIN	EGG COLOR	STATUS
Ornamental	Soft feather light	Europe	White	Fairly rare
COMB TYPE	FEATHER COLORING	BROODY	NO. OF EGGS/YEAR	BREED STANDARD
V-shaped	Chamois Golden Silver White White-crested black, and others	No	Very variable	APA PCGB Europe

White-crested black
Polish rooster

White-crested black
Polish hen

Polish day-old chick—the
crest is already visible

Rhode Island Red

Exemplary beginners' backyard bird—the best of both worlds

The quintessential hen—state bird of Rhode Island—and arguably the most successful breed ever created. What most people consider a traditional hen was created by crossing Asiatic fowls such as Shanghai, Malay, and Java to create an excellent dual-purpose breed with a reputation for reliable egg laying and fine carcass, along with a certain resistance to disease.

As with many breeds, the roosters can be aggressive, but as a rule the hens are calm, easy to handle, and will live happily in a large run or free range. They are not spectacular fliers so can be contained by a low fence.

Rhode Island Red hen

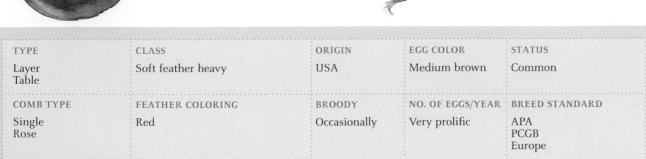

TYPE	CLASS	ORIGIN	EGG COLOR	STATUS
Layer Table	Soft feather heavy	USA	Medium brown	Common
COMB TYPE	FEATHER COLORING	BROODY	NO. OF EGGS/YEAR	BREED STANDARD
Single Rose	Red	Occasionally	Very prolific	APA PCGB Europe

Rhode Island Red rooster

Rhode Island Red
day-old chick

Rhode Island White

A handsome all-around bird with an easygoing character

This is not just a white Rhode Island Red but a separate breed altogether and does not enjoy the popularity of the former. It is made up of a mix of Wyandottes, Cochins, and Leghorns and was created around 1900.

When crossed with a Rhode Island Red, it produces the classic sex-link, that is, the female chicks are red and the males white. It is a calm, hardy bird with an easygoing character, which matures early, is an excellent all-arounder and would make a great choice as a slightly more unusual backyard bird.

Rhode Island White rooster

TYPE	CLASS	ORIGIN	EGG COLOR	STATUS
Layer Table	Soft feather heavy	USA	Medium brown	Common in USA Rare elsewhere
COMB TYPE	FEATHER COLORING	BROODY	NO. OF EGGS/YEAR	BREED STANDARD
Rose	White	Occasionally	Very prolific	APA

Rosecomb

King of the show ring and handsome to boot

Rosecombs are true bantams, that is, they have no large equivalent. Their origin is unclear but they have been found in the British Isles since the 15th century. They are smart little birds, with their rose combs that have a long leader or spike and large white earlobes.

Laying is not their strong point—their eggs are exceptionally small—and they are bred mainly as pets. In the breeding season, roosters can be aggressive, although the hens are friendly and easy to tame. They take kindly to confinement, although they love to forage and if free range it must be remembered that they are accomplished fliers. Thanks to a certain amount of inbreeding, they can suffer from infertility, which makes breeding difficult.

Rosecomb rooster with long leader or spike on comb

Rosecomb rooster

Rosecomb day-old chick

TYPE	CLASS	ORIGIN	EGG COLOR	STATUS
Ornamental	True bantam	UK	White or pale cream	Common
COMB TYPE	FEATHER COLORING	BROODY	NO. OF EGGS/YEAR	BREED STANDARD
Rose	Black and other variations	Occasionally	Very few	PCGB Europe

Rumpless Game

Characterful conversation piece that lacks a tail

In common with other rumpless breeds, the Rumpless Game lacks a parson's nose or uropygium, which is the fleshy protrusion at the end of the backbone from which the tail feathers grow. This was probably the result of a genetic mishap that occurred several hundred years ago, as this is an ancient breed. It is popular on the Isle of Man in the Irish Sea, where they specialize in tailless creatures, and known there as the Manx Rumpy.

Most commonly seen as bantams, this is a characterful breed, whose fairly upright stance and lack of tail give it a somewhat comical appearance. Though aggressive and noisy, it can be tamed and makes an interesting pet. Although known to have fertility problems, the hens make good mothers when they occasionally go broody.

Rumpless Game rooster

Rumpless Game hen

TYPE	CLASS	ORIGIN	EGG COLOR	STATUS
Game	Hard feather	UK	Tinted	Rare
COMB TYPE	FEATHER COLORING	BROODY	NO. OF EGGS/YEAR	BREED STANDARD
Small single	Many varieties	Occasionally	Few	PCGB Europe

Satsumadori

Glorious looks belie a belligerent character

These handsome birds were developed in Kagoshima, which was once known as Satsuma, on the island of Kyushu, Japan. They were bred originally for fighting with steel spurs, and like all Japanese game their small combs helped to prevent unnecessary injury.

Their proud bearing and piercing pale eye gives them a ferocious appearance. Unusually their magnificent tails fan out sideways, somewhat like a peacock's. The fact that they were originally fighting birds indicates that the roosters will be aggressive to other creatures, and in the case of the Satsumadori this includes humans.

satsumadori hen

silver duckwing
satsumadori rooster

TYPE	CLASS		ORIGIN	EGG COLOR	STATUS
Game Ornamental	Asian hard feather		Japan	White or tinted	Fairly rare
COMB TYPE	FEATHER COLORING		BROODY	NO. OF EGGS/YEAR	BREED STANDARD
Pea or triple	Black Black red Silver and gold duckwing White		Yes	Few	PCGB Europe

silver duckwing
satsumadori hen

Black red satsumadori
rooster

Scots Dumpy

An endearing personality with a mothering instinct

The exceptional feature of this breed is its very short legs, giving rise to a long list of other names, such as creepers and crawlers. It is an ancient breed and supposedly alerted the Scots and Picts to potential attacks from the Romans.

By the middle of the 20th century, the breed had nearly died out, but a pure line was found in Kenya, taken there by Lady Violet Carnegie in 1902, and reimported to give the existing stock a boost.

Hens of this breed make very good broodys; they were apparently particularly favored by gamekeepers to rear pheasants, but their own chicks can easily chill on wet grass, being so close to the ground. As short legs are a feature of this breed, the occasional chick that develops long legs should not be included in breeding stock.

Scots Dumpy hen

TYPE	CLASS	ORIGIN	EGG COLOR	STATUS
Layer Table	Soft feather light	UK	White	Common
COMB TYPE	FEATHER COLORING	BROODY	NO. OF EGGS/YEAR	BREED STANDARD
Single	Cuckoo most common but many other variations including black and white	Yes	Prolific	PCGB Europe

Scots Grey

A hardy bird well suited to its native climate

As its name suggests, this breed is found mostly in Scotland, where it has been in existence for over 200 years. It is a hardy bird that thrives in even the harshest climate, with the reputation of maturing quickly—an advantage in a country where the days are short in winter.

The Scots Grey has a fairly upright stance and long legs, supposedly as a result of some gamefowl in its ancestry. It is not an easy bird to tame and will roost up trees if given the chance—this is not a breed that will take kindly to close confinement. Although classed as a non-sitter, hens will occasionally go broody and when they do make excellent protective mothers.

scots Grey hen

TYPE	CLASS	ORIGIN	EGG COLOR	STATUS
Layer	Soft feather light	UK	White	Common
COMB TYPE	**FEATHER COLORING**	**BROODY**	**NO. OF EGGS/YEAR**	**BREED STANDARD**
Single	Cuckoo	Occasionally	Moderate	PCGB

Serama

Pocket-sized poultry that will enchant its owner

The Serama is the smallest breed of poultry in the world, which originally came from Malaysia and was thought to have been kept by Thai royalty as long ago as the 17th century. These tiny birds are even kept as house pets in Malaysia. They only recently arrived in the UK but already have a growing following—perhaps because their diminutive size makes them easy to keep in even the smallest space. They also possess a quiet crow, which will endear them to neighbors.

There are disadvantages to being small, in that if left to free range they are vulnerable not only to foxes but also to cats and birds such as magpies, hawks, and buzzards. They will also need to have warm housing—keeping a larger bantam such as a Silkie will provide natural heating. Their eggs are equally tiny, being not much larger than a quail egg.

The birds carry their tails at 90° with their chests protruding and give a proud appearance. Roosters may weigh as little as 12 oz. and be no more than 6 in. high. They are divided into classes by weight:

Class A	rooster 12 oz.
	hen 11 oz.
Class B	rooster 17 oz.
	hen 15 oz.
Class C	rooster 20 oz.
	hen 17 oz.

serama hen

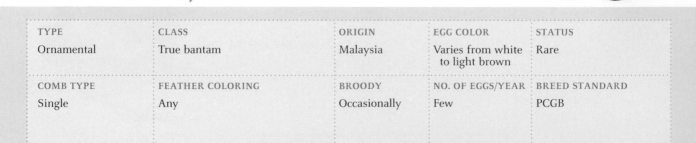

TYPE	CLASS	ORIGIN	EGG COLOR	STATUS
Ornamental	True bantam	Malaysia	Varies from white to light brown	Rare
COMB TYPE	**FEATHER COLORING**	**BROODY**	**NO. OF EGGS/YEAR**	**BREED STANDARD**
Single	Any	Occasionally	Few	PCGB

Serama rooster

Serama
day-old chick

Shamo

Ferocious fighter with personality but a quiet voice

This is a breed that comes in a variety of sizes: O-Shamo, meaning large, and Chu-Shamo, meaning medium. The O-Shamo is the tallest of the Japanese game and can stand at up to 30 in in height. This is a very ancient breed and was initially imported into the United States in 1874 specifically for cockfighting; it didn't reach the UK or Europe for another 100 years.

Its very upright stance, heavy overhanging brows, bare face and neck, and patches of bare red skin down the keel give it a very ferocious demeanor, and indeed it is an extremely aggressive breed. The chicks may even fight each other from the moment they leave the shell and have to be separated. Although antagonistic to each other, these birds seem to like humans and are easy to tame, making characterful and charming pets, with the added bonus that they tend not to crow as much as some.

shamo hen

TYPE	CLASS		ORIGIN	EGG COLOR	STATUS
Game	Asian hard feather		Japan	White or tinted	Rare in USA Fairly common in UK and Europe
COMB TYPE	FEATHER COLORING		BROODY	NO. OF EGGS/YEAR	BREED STANDARD
Pea or triple	Black-breasted red with wheaten hen, and many others		Yes	Very few	APA PCGB Europe

shamo day-old chick

shamo rooster

Sicilian Buttercup

A pretty princess with a crown on her head

This is a very old breed, which may have been known even in biblical times—it was certainly around in Italy in the 16th century, when it was frequently depicted by Italian artists of that time. The first birds were imported into the USA in 1835, where they became known just as Buttercups or Butters, and did not make it to British shores until around 1913. Originally all birds had red earlobes, but the American standard now requires a white one—birds imported more recently tend to be white and the British standard now states that lobes need not be solid red but more than one-third white will count as a defect.

The outstanding feature of this breed is the enormous cup-shaped comb, which is actually two single combs joined at the front and back; this takes time to reach its final splendor but its size makes it prone to frostbite. The cockerels have the normal farmyard bird coloring but the hens are a beautiful golden color, with black spangles and black tail—there is also a silver version, where the ground color is silver as opposed to gold.

This is an early-maturing breed—cockerels can crow at one month—but the birds do tend to be fairly flighty and resist confinement.

Sicilian Buttercup rooster with large cup-shaped comb

TYPE	CLASS	ORIGIN	EGG COLOR	STATUS
Layer	Soft feather light	Italy	White	Rare
COMB TYPE	FEATHER COLORING	BROODY	NO. OF EGGS/YEAR	BREED STANDARD
Cup	Gold Silver	No	Moderate	APA PCGB Europe

Gold Sicilian
Buttercup hen

Sicilian Buttercup
day-old chicks

Silkie

This is the powder-puff of the poultry world

This is a very ancient breed, originally from Asia though its exact origins are unknown. Marco Polo brought back stories of these strange fowl when he returned from his travels in the 13th century.

Silkies are totally different from most other poultry, having a melanotic gene, which results in their skin, and even their bones, being black. Their very odd feathers are fluffy and more like fur but do not retain heat or provide waterproofing in the same way as normal feathers, and this causes them to dislike getting wet. The Silkie is also one of the few breeds with a fifth toe and completes its unique look with a topknot and sometimes even a muff.

They make excellent pets, being docile and friendly, but do spend a good deal of their time broody and are widely used as surrogate mothers for all kinds of other birds.

Although small they are not considered to be bantams but classed as light soft feather.

silkie hen

TYPE	CLASS	ORIGIN	EGG COLOR	STATUS
Layer	Soft feather light	Asia	Pale cream or tinted	Common
COMB TYPE	FEATHER COLORING	BROODY	NO. OF EGGS/YEAR	BREED STANDARD
Walnut or cushion	Many varieties	Yes	Few	APA PCGB Europe

Dark silkie hen

silkie day-old chick

silkie hen

Spanish

An unusual but elegant choice with a unique charm

The white-faced Spanish is one of the oldest of the Mediterranean breeds, having been known since the 16th century, or even before. It is also one of the oldest known in the United States, having been brought there via the Caribbean by early Spanish settlers and made an appearance at the first poultry show in Boston in 1849.

In the hen the comb droops to one side, although it should remain upright in the roosters, whose large white faces give the birds a somewhat haughty look. They are, in fact, fairly noisy and not always friendly creatures, though with a charm of their own, which do not take kindly to being over-contained, preferring free range. This breed is somewhat slow to mature, the white face taking a year or so to develop, but it will eventually reward you with a good number of large white eggs.

spanish rooster

spanish hen

spanish day-old chick

TYPE	CLASS	ORIGIN	EGG COLOR	STATUS
Layer	Soft feather light	Mediterranean	White	Rare
COMB TYPE	FEATHER COLORING	BROODY	NO. OF EGGS/YEAR	BREED STANDARD
Single	Black	No	Moderate	APA PCGB Europe

Sulmtaler

Delicatessen from Austria that thrives on maize

Named after the Sulm valley in the region of Stiermarken in south-western Austria, the Sulmtaler was developed in the 19th century to fill the need for large, heavy hens. What was produced was a hardy, fast-growing breed that was easy to fatten, particularly if fed maize. It soon spread across Europe, and the Germans produced a bantam version after the Great War.

The hens have fairly sturdy bodies, with twisted single combs and a tuft-like crest on the back of the head. Although calm natured, these birds are good fliers and will require reasonably high fencing to contain them. They are classed as rare in Britain but are still plentiful in Austria and other European countries.

Wheaten sulmtaler hen

TYPE	CLASS	ORIGIN	EGG COLOR	STATUS
Layer Table	Soft feather heavy	Austria	Pale cream or light brown	Rare
COMB TYPE	FEATHER COLORING	BROODY	NO. OF EGGS/YEAR	BREED STANDARD
Single twisted	White Wheaten	Occasionally	Prolific	PCGB Europe

Sultan

An unusual-looking breed that make charming pets

Sultans were introduced in the 19th century from Turkey, where they were called Serai-Tavuk, which means "fowls of the Sultan", and kept as pets by Turkish royalty. Those found in Britain all originated from five that were imported by Mrs. Elizabeth Watts in January 1854.

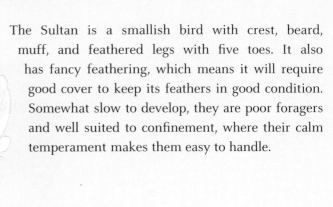

The Sultan is a smallish bird with crest, beard, muff, and feathered legs with five toes. It also has fancy feathering, which means it will require good cover to keep its feathers in good condition. Somewhat slow to develop, they are poor foragers and well suited to confinement, where their calm temperament makes them easy to handle.

sultan hen

TYPE	CLASS	ORIGIN	EGG COLOR	STATUS
Ornamental	Soft feather light	Turkey	White	Rare
COMB TYPE	FEATHER COLORING	BROODY	NO. OF EGGS/YEAR	BREED STANDARD
V-shaped	White	No	Very few	PCGB Europe

sultan rooster

sultan
day-old chick

Sumatra

A glamorous breed with glistening plumage

As its name suggests, this is an ancient breed from the island of Sumatra. It is a very active breed and intolerant of confinement, which means it will need very high fencing even if its wings are clipped, as it can also jump high. Pens may even need to be covered to contain it, and the house will also need to be larger than normal to accommodate the long tail.

Sumatra rooster

Sumatras have wonderful lustrous, shiny feathers, changing color in sunlight. Their long tails flow horizontally behind, fairly similar to a pheasant, and altogether they look graceful and exotic. They have plum-colored faces, a coloring sometimes called "gypsy", and uniquely some have several spurs on each leg. This is not an aggressive breed as far as gamefowl go, but cockerels will fight, particularly in the breeding season, which with this breed is slightly later in the year than most.

TYPE	CLASS	ORIGIN	EGG COLOR	STATUS
Game Ornamental	Soft feather light	Asia	White	Rare

COMB TYPE	FEATHER COLORING	BROODY	NO. OF EGGS/YEAR	BREED STANDARD
Pea or triple	Black Blue White	Yes	Moderate	PCGB Europe

An attractive, popular, and reliable breed

This is one of Britain's oldest breeds and also one of the commonest—the light, with its black neck and tail and white body, being the most popular. Originally there was a black–red (now called brown) version, which was well camouflaged when sitting; but a "light", really Columbian, was developed, as the white feather stubs didn't show on the carcass when plucked. The light was also useful for sex-linkage; typically a light Sussex hen mated with a Rhode Island Red rooster produced chicks with dark females and pale males.

The speckled Sussex has mahogany feathers with black-and-white mottling that glows green in the sun, and there is also a buff-and-silver version.

All Sussex hens are hardy, adaptable, and easily handled, in fact the perfect backyard bird.

speckled sussex feather

Light sussex hen

TYPE	CLASS	ORIGIN	EGG COLOR	STATUS
Layer Table	Soft feather heavy	UK	Light brown	Common
COMB TYPE	FEATHER COLORING	BROODY	NO. OF EGGS/YEAR	BREED STANDARD
Single	Brown Buff Light Silver Speckled	Yes	Very prolific	APA PCGB Europe

Light sussex rooster

speckled sussex hen

Thai Game

A ferocious member of the Asian hard feathers

Although a large bird related to the Malay, the Thai Game is lighter than the Shamo, with a higher tail carriage. It was originally developed in Thailand for cockfighting—a sport that although banned in the Western world still continues in the East.

Thais are found in many colors but red black is the most common.

Red black Thai Game rooster

TYPE	CLASS	ORIGIN	EGG COLOR	STATUS
Game	Asian hard feather	Thailand	White or tinted	Fairly common
COMB TYPE	FEATHER COLORING	BROODY	NO OF EGGS/YEAR	BREED STANDARD
Pea or triple Walnut or cushion	Any	Yes	Few	PCGB Europe

Thüringian

A sadly rare breed with charm and a full beard

There have been records of Thüringian hens in Germany since 1793, but they have only been seen in Britain since 2000. They are charming birds with a docile temperament, making them ideal for small runs, and with their full beards and frequently spangled feathers, they are unusual as well.

Thüringian hen

Silver spangled Thüringian hen

TYPE	CLASS	ORIGIN	EGG COLOR	STATUS
Ornamental	Soft feather light	Germany	White	Rare
COMB TYPE	**FEATHER COLORING**	**BROODY**	**NO. OF EGGS/YEAR**	**BREED STANDARD**
Single	Black Chamois, gold and silver spangled	Occasionally	Moderate	PCGB Europe

Transylvanian Naked Neck

A conversation piece that is also a reliable layer

Transylvania is now a region of Romania and not necessarily where this breed originated, as they are found in many other parts of Europe and the Middle East. Also known as a Turken because it faintly resembles a turkey, the Naked Neck has to be one of the strangest, some might even say ugliest, breeds of poultry. The lack of feathers on the neck is caused by a gene that reduces the size and density of the plumage. This is an advantage in a hot climate (although they can suffer sunburn on their naked necks), but these mild-mannered, friendly birds are hardy and tolerant of cold. They are also thought to be exceptionally immune to disease.

The fact that they are slow to mature also means that they produce good meaty carcasses, which are quick to pluck due to the thinness of the plumage. This lack of plumage also has the advantage that, rather than putting protein toward producing feathers, it goes into egg production, and they are good, reliable layers of tinted eggs.

Transylvanian Naked Neck hen

TYPE	CLASS	ORIGIN	EGG COLOR	STATUS
Layer Table	Soft feather heavy	Eastern Europe	Tinted	Rare
COMB TYPE	FEATHER COLORING	BROODY	NO. OF EGGS/YEAR	BREED STANDARD
Single	Black Blue Buff Cuckoo Red White	Occasionally	Prolific	APA PCGB

Transylvanian
Naked Neck hen

Transylvanian Naked Neck
day-old chick—its naked
neck already visible

Tuzo

A slightly sinister-looking Asian hard feather

The Tuzo was probably developed from the Nankin Shamo, a similar true bantam now very rare. Folklore has it that it was kept by Japanese nobility, but there is no record of this in Japan—it does, however, descend from Asian bloodlines.

An elegant, small bird with proud bearing and shiny black plumage tight to the body. Some lines have black spurs and, it is said, black tongues as well, though this may also be folklore.

Tuzo rooster with upright stance and proud bearing

TYPE	CLASS	ORIGIN	EGG COLOR	STATUS
Game	Asian hard feather True bantam	USA and Europe	White	Rare
COMB TYPE	**FEATHER COLORING**	**BROODY**	**NO. OF EGGS/YEAR**	**BREED STANDARD**
Pea or triple	Black	Yes	Few	PCGB Europe

Vikinghen

A free-ranging hen with personality and a variety of coloring

When the Vikings reached Iceland, they brought their poultry with them, and astonishingly for 1,000 years the Vikinghen, or Landnamshaena as it is known in Iceland, remained the only breed found there. In the 1930s, brown and white Leghorns were introduced and soon outdid the Vikinghens as layers and meat producers, and numbers dwindled. The formation of a breed club regenerated interest, and the Vikinghen is now found in small numbers in the United States and several European countries.

This ancient breed comes in a wide range of colors and comb types, and may even have head-tufts. They are tough, self-sufficient, and self-possessed and would not be happy in a small run. Even so they are exceptionally friendly and make good pets.

Vikinghen

Vikinghen day-old chicks

TYPE	CLASS	ORIGIN	EGG COLOR	STATUS
Layer	Soft feather light	Iceland	White to light brown	Rare
COMB TYPE	**FEATHER COLORING**	**BROODY**	**NO. OF EGGS/YEAR**	**BREED STANDARD**
Rose Single	Many varieties	Yes	Very prolific	None

Vorwerk

A well-dressed bird for the backyarder to enjoy

These striking hens were developed in Hamburg in 1900, with the aim of producing an economical utility breed. What resulted was an ideal small farmers' bird—quick to mature, good forager, and prolific egg production that continued through the winter.

Their striking plumage is fairly similar to Lakenvelders, though in a different colorway, being black neck and tail and chestnut body, and they are classed as the same breed by the APA, although the Vorwerk bantam is recognized. They are tolerant birds that are happy confined as well as free range, and the males have the useful trait that they tolerate each other more than most breeds. The hens are not prone to frequent broodiness, but when they do hatch, the chicks are exceptionally lively.

Vorwerk hen

Vorwerk day-old chick

TYPE	CLASS	ORIGIN	EGG COLOR	STATUS
Layer	Soft feather light	Germany	White or pale cream	Rare
COMB TYPE	FEATHER COLORING	BROODY	NO. OF EGGS/YEAR	BREED STANDARD
Single	Black neck and tail, buff body	Occasionally	Very prolific	PCGB Europe

Welsummer

An all-around beautiful breed, ideal for the backyarder

This is a fairly modern breed, created by a farmer in Welsum, Holland at the end of the 19th century, who crossed various breeds, including Barnevelders, until he found the stable cross he wanted. It didn't reach British shores until the late 1920s and is still not common in the United States.

Welsummer rooster

Apart from being an exceptionally pretty hen and handsome cockerel of a traditional farmyard type, its chief claim to fame is the wonderful dark, sometimes speckled, brown of its eggs. This hardy breed is generally adaptable; it is economical to feed, being an excellent forager, though occasional strains do tend to be on the flighty side. The hen has lovely partridge coloring, enhanced by the light shafts of the feathers. Hens will go broody but not excessively so, making reliable mothers, and when hatched an experienced eye can distinguish hens from cockerels by the color of the stripe on their heads— the cockerel being darker. This does take practice and needs comparison, but by six weeks or so the cockerels will become obvious as their combs start to appear.

TYPE	CLASS	ORIGIN	EGG COLOR	STATUS
Layer Table	Soft feather light	Holland	Dark brown speckled	Common
COMB TYPE	FEATHER COLORING	BROODY	NO. OF EGGS/YEAR	BREED STANDARD
Single	Partridge Silver duckwing White	Occasionally	Prolific	APA PCGB Europe

Red partridge
Welsummer hen

Welsummer day-old chicks

Wyandotte

A superb and showy dual-purpose bird with a strong character to match

This breed acquired its name from a Native American tribe called the Wendat, which became corrupted to Wyandotte by settlers when it was developed in the late 19th century.

Wyandottes come in a wide range of colors, the silver laced being the original and perhaps most striking variety. They have a somewhat rotund appearance, short tail, sturdy legs, and slightly loose feathers, which makes them look fluffy. This is a robust breed, the hens of which make excellent mothers. Most birds are calm and docile, but their strong characters can sometimes make the occasional one seem unfriendly—this is, however, an ideal beginners' breed.

Wyandotte rooster

Gold laced, blue laced, and silver laced

TYPE	CLASS	ORIGIN	EGG COLOR	STATUS
Layer Table	Soft feather heavy	USA	Light to medium brown	Common
COMB TYPE	**FEATHER COLORING**	**BROODY**	**NO. OF EGGS/YEAR**	**BREED STANDARD**
Rose	Large variety including laced and penciled	Yes	Very prolific	APA PCGB Europe

Blue laced
Wyandotte rooster

silver laced
Wyandotte hen

Wyandotte
day-old chick

Yakido

A miniature Asian hard feather that only comes in black

This is a true bantam variety of Shamo, created in the 1850s in Mie Province on the main Japanese island of Honshu. It has an upright stance with a fierce, proud bearing and was originally created as a sparring partner for the larger O-Shamo.

Unlike many hard feather breeds, the Yakido only comes with black plumage. It has a bright red triple pea comb and yellow, or yellow with black, spotted legs. As with all gamefowl, these birds are best kept as pairs or trios, that is, one rooster and two hens. No new birds can be introduced, as they simply will not be tolerated.

Yakido rooster—with a fierce, proud bearing

TYPE	CLASS	ORIGIN	EGG COLOR	STATUS
Game	Asian hard feather True bantam	Japan	White or tinted	Common
COMB TYPE	FEATHER COLORING	BROODY	NO. OF EGGS/YEAR	BREED STANDARD
Pea or triple	Black	Yes	Very few	PCGB Europe

Yamato-Gunkei

With the character and temperament to overcome its looks

An ancient ornamental breed and largest of the small Shamos, the Yamato-Gunkei has a fairly large frame with chunky legs. Its somewhat outsize head has a pendulous dewlap, which is very wrinkled and becomes more so with age, which can result in the bird looking quite grotesque—they are sometimes likened to the chicken version of a bonsai tree. The feathers are hard and sparse, leaving the breastbone completely bare, and the short shrimp-tail should point between the legs.

The wings are held away from the body and the shoulder coverts should show clearly on the back giving what is known in poultry circles as the "five hills" outline, that is, wing/shoulder covert/back/shoulder covert/wing.

It gives a tough appearance but is, in fact, quite delicate and needs good housing.

Yamato-Gunkei rooster

TYPE	CLASS	ORIGIN	EGG COLOR	STATUS
Game	Asian hard feather	Japan	White or tinted	Common
COMB TYPE	FEATHER COLORING	BROODY	NO. OF EGGS/YEAR	BREED STANDARD
Pea or triple Walnut or cushion	Many varieties	Yes	Very few	PCGB Europe

Yamato-Gunkei hen

Yamato-Gunkei rooster

Yokohama

A beautiful breed with a remarkably long tail

Originally documented in China between 600 and 800 AD, modern-day Yokohamas, in particular the red-saddled white, were developed in Germany in the 19th century from various Japanese long-tailed breeds.

Their very fancy feathering needs good protection and a large house to accommodate their exceptionally long tails. This elegant, docile breed is slow to mature but the hens make good, protective mothers. The male bird's tail can grow up to 3 ft. a year and they also have extra-long saddle feathers.

In the UK, many of the Japanese long-tailed and long-crower breeds are included in the Yokohama standard.

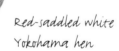

Red-saddled white Yokohama hen

TYPE	CLASS	ORIGIN	EGG COLOR	STATUS
Ornamental	Soft feather light	Japan and Europe	White or tinted	Rare
COMB TYPE	FEATHER COLORING	BROODY	NO. OF EGGS/YEAR	BREED STANDARD
Pea or triple Walnut or cushion	Many varieties	Occasionally	Few	PCGB Europe

Common ailments

PROBLEM	SIGNS	ACTION
PARASITES		
Fleas	Birds may scratch and create bare patches – poor laying—look unwell	Treat with flea powder, including in house. Increase hygiene
Lice	Similar to above—check vent for white clusters of eggs	As above
Mites: Northern fowl mite	Dirty-looking feathers on pale-colored birds. This mite lives on the bird	Increased hygiene in house. Flea powder or proprietary spray every 8 days
Red mite	This mite lives in the house, on perches, etc. – you will be itchy if you have been inside	As for above but cycle is 7–10 days
Worms	Hard to spot	Worm birds twice a year with a proprietary product
Scaly leg	Scales lifted on the leg—leg looks crusty and enlarged because of the tiny mite that burrows under the scales. Legs return to normal after next molt, when cured	Proprietary anti-scaly leg product or immerse leg for 20 seconds in a jar of rubbing alcohol once a week for 4 weeks
ILLNESSES		
Crop bound	Sour-crop—bulging soft crop when bird gets up in morning—hen appears ill—caused by fungus called Candida albicans	Feed live yogurt—encourage vomiting by holding bird's head down and massaging crop
	Impacted crop—full, hard crop when bird gets up—can be caused by eating long grass	As above—gently massage crop
Coccidiosis	Milky white diarrhea, occasionally with blood. 3–8-weeks-old danger period. Sudden death	Hatch as early as possible in the year—parasite lives on damp grass, spread in droppings.
		Good hygiene. Vaccine available. Chick crumbs and pellets contain anti-coccidiostats
Colds	Runny nose and coughing	Proprietary preparation from vet or feed merchant
Marek's disease	Many and varied—dead birds, paralysis, not growing, unnaturally hungry	Vaccination at day old. Never mix young with old birds

PROBLEM	SIGNS	ACTION
ILLNESSES		
Newcastle disease (fowl pest)	Notifiable. Twisted necks, birds floppy and unsteady, breathing difficulty. Death	Vaccination—if diagnosed, mandatory slaughter of flock
Avian influenza	H5N1 breeds in respiratory and intestinal tracts and is transmitted bird to bird. Difficulty breathing and sudden death	Hygiene—covering run to keep out wild birds during outbreak. Vaccination
OTHER PROBLEMS		
Fishy eggs	Eggs have fishy flavor	Found occasionally, no known cause or cure
Egg eating	Birds start eating eggs in nesting boxes	Starts with broken egg—keep supply of shell available. Collect eggs frequently. Make nesting boxes dark
Feather pecking	Hens developing bare, sore patches on neck, rump, and vent	Less overcrowding—outdoor perches and dust baths to combat boredom

Culling

There will come a time when one of your flock may need to be put out of its misery—possibly it will have been attacked by a dog or become seriously ill—and it is your responsibility to do this. If you hatch your own chicks, you must be aware that 50% will most likely be cockerels, which you may not even be able to give away, let alone sell. The obvious answer is to rear them for the table and end their lives in the most humane way possible.

You could take them to the vet or find another poultry keeper nearby who will do the job for you—your local poultry society could help with this. There are various humane dispatchers available, but it is generally thought, and recommended by the Humane Slaughter Association in the UK, that neck dislocation is the kindest and swiftest method. Before attempting it yourself, it would be a good idea to find someone to give you a demonstration.

How to wring a neck

If possible collect the bird for slaughter after dark when it is quietly on its perch. Hold the legs firmly in one hand, and with the bird's body against your thigh and head hanging down, place the index and middle fingers (or thumb and index finger if you prefer) on each side of the neck with the head in the palm of your hand. Bend the head slightly outward and give a sharp, twisting pull downward, at the same time bending the neck backward. This dislocates the neck from the head and death follows instantly. There will almost always be a certain amount of flapping, but this is simply post-mortal nerves and will cease shortly.

Breed classifications

SOFT FEATHER HEAVY

Australorp
Barnevelder
Brahma
Buckeye
Chantecler
Cochin
Crevecoeur
Croad Langshan
Delaware
Dominique
Dorking
Faverolles
German Langshan
Houdan
Ixworth
Jersey Giant
La Bresse / Blue Foot Chicken
La Fleche
Lincolnshire Buff
Malines
Marans
Modern Langshan
New Hampshire Red
Norfolk Grey
Orloff
Orpington
Plymouth Rock
Rhode Island Red
Rhode Island White
Sulmtaler
Sussex
Transylvanian Naked Neck
Wyandotte

SOFT FEATHER LIGHT

Ancona
Andalusian
Appenzeller
Araucana
Ardenner
Brabançonne
Brabanter
Braekel
Breda Fowl
Campine
Catalana
Cream Legbar
Dandarawi
Derbyshire Redcap
Fayoumi
Friesian
Hamburg
Java
Kraienköppe
Kurokashiwa
Lakenvelder
Leghorn
Marsh Daisy
Minorca
Old English Pheasant Fowl
Penedesenca
Phoenix
Polish
Scots Dumpy
Scots Grey
Sicilian Buttercup
Silkie
Spanish
Sultan
Sumatra
Thüringian
Vikinghen
Vorwerk
Welsummer
Yokohama

HARD FEATHER

Cubalaya
Indian / Cornish Game
Modern Game
Old English Game (Carlisle)
Old English Game (Oxford)
Rumpless Game

ASIAN HARD FEATHER

Aseel (Asil or Reza Asil)
Ko Shamo
Kulang
Malay
Satsumadori
Shamo
Thai Game
Tuzo
Yakido
Yamato Gunkei

TRUE BANTAM

Barbu d'Anvers (Antwerp Belgian)
Barbu d'Uccle (Millefleur)
Barbu du Watermael
Booted
Burmese
Japanese
Ohiki
Pekin
Rosecomb
Sebright
Serama

Quick reference guide

	TYPE			CLASS		EGG COLOR					BROODINESS		
	layer	table	ornamental	bantam	game	blue	white	light brown tinted	medium brown	dark brown	non-sitter	occassionally	often
ANCONA	●						●				●		
ANDALUSIAN	●						●				●		
APPENZELLER	●		●				●					●	
ARAUCANA	●					●						●	
ARDENNER	●						●					●	
ASEEL					●			●					●
AUSTRALORP	●	●						●					●
BARBU D'ANVERS			●	●			●						●
BARBU D'UCCLE			●	●			●						●
BARBU DU WATERMAEL			●	●				●					●
BARNEVELDER	●									●		●	
BOOTED BANTAM			●	●			●					●	
BRABANÇONNE	●						●				●		
BRABANTER	●	●					●				●		
BRAEKEL	●						●				●		
BRAHMA	●	●						●	●	●			●
BREDA FOWL	●	●					●				●		
BUCKEYE	●	●							●				●
BURMESE			●	●			●						
CAMPINE	●						●					●	
CATALANA	●	●						●			●		
CHANTECLER	●	●						●				●	
COCHIN	●	●						●					●
CREAM LEGBAR	●					●							●
CREVECOEUR	●	●					●				●		
CROAD LANGSHAN	●	●							●			●	
CUBALAYA	●	●			●		●						●
DANDARAWI	●							●				●	
DELAWARE	●	●							●			●	
DERBYSHIRE REDCAP	●	●					●				●		
DOMINIQUE	●	●							●			●	
DORKING	●	●					●						●

	TYPE			CLASS		EGG COLOR					BROODINESS		
	layer	table	ornamental	bantam	game	blue	white	light brown tinted	medium brown	dark brown	non-sitter	occassionally	often
FAVEROLLES	●	●						●				●	
FAYOUMI	●							●			●		
FRIESIAN	●						●					●	
GERMAN LANGSHAN	●	●							●			●	
HAMBURG	●						●					●	
HOUDAN	●	●	●				●					●	
INDIAN / CORNISH GAME		●			●			●					●
IXWORTH	●	●						●					●
JAPANESE BANTAM			●	●			●	●	●	●			●
JAVA	●	●							●				●
JERSEY GIANT	●	●							●				●
KO SHAMO				●	●			●					●
KRAIENKÖPPE	●							●				●	
KULANG					●			●					●
KUROKASHIWA			●		●			●					●
LA BRESSE / BLUE FOOT	●	●					●				●		
LA FLECHE	●	●					●				●		
LAKENVELDER	●						●	●			●		
LEGHORN	●						●				●		
LINCOLNSHIRE BUFF	●	●						●				●	
MALAY					●			●				●	
MALINES	●	●						●				●	
MARANS	●									●			●
MARSH DAISY	●							●				●	
MINORCA	●						●				●		
MODERN GAME					●		●	●					●
MODERN LANGSHAN	●								●				●
NEW HAMPSHIRE RED	●	●						●	●				●
NORFOLK GREY	●	●					●						●
OHIKI			●	●				●					●
OLD ENGLISH GAME					●			●					●
OLD ENGLISH PHEASANT FOWL	●	●					●	●			●		
ORLOFF		●	●					●			●		
ORPINGTON	●	●							●				●
PEKIN			●	●			●	●					●
PENEDESENCA	●									●			

	TYPE			CLASS			EGG COLOR				BROODINESS		
	layer	table	ornamental	bantam	game	blue	white	light brown tinted	medium brown	dark brown	non-sitter	occassionally	often
PHOENIX			●				●					●	
PLYMOUTH ROCK	●	●						●				●	
POLISH			●				●				●		
RHODE ISLAND RED	●	●							●			●	
RHODE ISLAND WHITE	●	●							●			●	
ROSECOMB			●	●			●	●				●	
RUMPLESS GAME					●			●				●	
SATSUMADORI			●		●		●	●					●
SCOTS DUMPY	●	●					●						●
SCOTS GREY	●						●						●
SEBRIGHT			●	●			●	●				●	
SERAMA			●	●			●	●				●	
SHAMO					●		●	●					●
SICILIAN BUTTERCUP	●						●				●		
SILKIE	●						●						●
SPANISH	●						●				●		
SULMTALER	●	●					●					●	
SULTAN			●				●				●		
SUMATRA			●		●		●						●
SUSSEX	●	●						●					●
THAI GAME					●		●	●					●
THÜRINGIAN			●				●					●	
TRANSYLVANIAN NAKED NECK	●	●						●				●	
TUZO				●	●		●						●
VIKINGHEN	●						●	●					●
VORWERK	●						●	●				●	
WELSUMMER	●	●								●		●	
WYANDOTTE	●	●						●	●				●
YAKIDO				●	●		●	●					●
YAMATO-GUNKEI					●		●	●					●
YOKOHAMA			●				●	●				●	

Egg colors

White

Off-white

Pale cream

Tinted

Pale tinted pink

Yellow

Brownish-yellow

Pinkish-plum

Light brown

Medium brown

Reddish-brown

Dark brown

Dark brown speckled

Blue

Khaki

Useful websites

www.amerpoultryassn.com
The site of the American Poultry Association "to promote and protect the standard-bred poultry in all its phases".

www.bantamclub.com
Site of the American Bantam Club, a "national organization to promote breeding and exhibiting of all kinds of bantams".

www.usda.gov
The site of the U.S. Department of Agriculture—the department responsible for developing and executing U. S. federal government policy on farming, agriculture, and food.

www.fws.gov
The site of the U.S. Fish & Wildlife Service—the federal government agency, within the Department of the Interior, dedicated to the management of fish, wildlife, and habitats.

www.feathersite.com
Comprehensive poultry site with lots of photos and information.

www.poultryclub.org
The site of the Poultry Club of Great Britain—it gives lists of breed clubs and societies with their secretaries and phone numbers. Also a great deal of useful advice from breed lists to care, etc.

www.rarepoultrysociety.co.uk
This U.K. society was formed to cater for the rare and endangered breeds that do not have their own breed club or society—it includes new breeds from countries other than the U.K.

www.rbst.org.uk
The Rare Breeds Survival Trust is a U.K. charity founded to protect all native breeds of farm animals, including poultry.

Acknowledgements

I would like to thank the following people, who very kindly allowed me to use their chickens as models for my illustrations or supplied me with photos:

Andalusian rooster and hen heads; Campine (gold rooster): Ariel Redmond

Appenzeller; Campine rooster; Marsh Daisy rooster, hen, and head; Rhode Island White rooster: Chris Graham, Editor *Practical Poultry*

Araucana hen: Anne Cushing

Araucana (lavender rooster); Friesian hen, rooster, and chick: Jeni Stanton

Ardenner rooster; Barbu d'Anvers; Barbu d'Uccle rooster and hen; Barbu du Watermael; Brabançonne head; Breda Fowl hen; Brakel hen; Cornish Game hen and rooster; Japanese Bantam rooster; Leghorn (black hen and head, brown hen); Malines hen; New Hampshire hen; Orloff head; Rosecomb head; Sebright hen and rooster; and several chicks: Johan Opsomer

Ardenner hen and heads: Stuart Sutton

Barnevelder rooster and hen: Ian MacDonald

Brabanter chick: Lisa West

Brabanter hen: Mary Wilson

Brahma rooster: Sylvia Brown

Brahma head: Joachim Dippold

Brahma (dark hen); Jersey Giant hen; Orloff rooster (photo: Henny Tate); Pekin hen; Satsumadori hen; Scots Grey hen (photo: Henny Tate); Silkie hen; Sultan hen: Maurice

Dukes

Buckeye hen; Java hen; Ko Shamo rooster; Norfolk Grey hen: Tony Beardsmore

Burmese rooster; Modern Langshan rooster: The Cobthorn Trust

Chantecler hen: Yvonne Hillsden

Cochin black hen; Spanish (white-faced rooster): Andy Marshall

Cream Legbar chicks: Rowena Evans

Crevecoeur rooster; La Fleche head: Jean-Claude Periquet

Crevecoeur hen: Rob Gibson

Croad Langshan rooster: The Croad Langshan Club

Dandarawi hen and rooster head: Dr Nabil Makled

Derbyshire Redcap hen: Christine Taylor

Dominique hen; Lakenvelder hen: Chalkhill poultry

Dorking chicks: Nicki Stannard

Faverolles rooster: Hazel Marks

Faverolles hen; Polish hen (chamois): David Brandreth

Fayoumi hen; Rumpless Game hen and rooster: Robert Stephenson

German Langshan rooster: Karen Pimlott

Hamburg gold-penciled hen: Matt Gavenlock

Ixworth hen; Vorwerk hen: Gillian Dixon

Ixworth chick; Vorwerk chick: Jo Seccombe

Japanese Bantam hen; Serama rooster; white Wyandotte head: Charlotte Carneigie

Java; Old English Game Carlisle type rooster; Plymouth Rock (barred hen); Polish rooster and head; and several chicks: Derek Sasaki of My Pet Chicken

Kulang rooster: Wilem van Ballekom

Kurakoshiwa rooster and hen head; Shamo hen: Julia Keeling

La Fleche hen; Kraienköppe hen; Scots Dumpy hen; Transylvannian Naked Neck hen: Peter Hayford; photos: Rinsey Mills

Leghorn (Exchequer rooster): Rod and Moira Attrill

Lincolnshire Buff rooster: Lucy Hampstead

Malay; Old English Game Oxford type rooster and head: Stuart and Jennifer Gamble

Marans rooster and hen: Derrie Watchorn

Modern Game rooster and hen: Jennifer O'Sullivan

Ohiki: Jaroslaw Mazur

Old English Game Carlisle type hen: Joan Barry

Old English Pheasant Fowl hen: Tracy Eden

Orpington (buff hen): Sue Lambert

Pekin (lavender rooster): Henrietta Fiddian-Green

Penedesenca hen: Amadeu Francesch

Serama hen: Mrs. J. Cable

Shamo rooster: John Benson

Sicilian Buttercup hen: Eric Coppinger; heads: Nüle Mersch

Spanish (white-faced hen): Tricia Coles

Sultan rooster: Emma Lewis

Sulumtaler hen: Georg Zohrer

Sumatra rooster: Wanda Zwart

Sussex (light hen): Molly Mahon

Sussex hen with chicks: Jane Burton

Sussex rooster: Danny and Carolyne Pecorelli

Thai rooster: Jim Zook

Thüringian hen: Liz Holt

Vikinghen: Johanna Haroardottir

Wyandotte (silver laced hen): owner: Allan Brookes; photo: Chris Graham

Wyandotte (blue laced rooster): Collette Roberts

Yamato-Gunkei rooster and hen: Dirk Henken

Yokohama hen: Fleur Swanton

I would also like to thank Anne Merriman of the Rare Poultry Society and Andrew Sheppy of The Cobthorn Trust for their help, and particularly Julia Keeling for her tireless advice on Asian hard feathers.

Glossary

auto-sexing: a breed in which male chicks are lighter than females

bantam: small version of large chicken

barring: stripes of two colors across a feather

beard: feathers in a small clump under beak, e.g., Faverolles

booted: having feathers on the legs and feet and including vulture hocks

cape: feathers between neck and shoulder

cock: UK term to describe male bird after its first molt

cockerel: male bird before its first molt, see "rooster"

columbian: plumage color where body is white, hackles are black with white lacing, and tail is mainly black

comb: fleshy protruberance on bird's head

crest: feathers on top of a bird's head, known as "top-knot" or, in an Old English Game, "tassel"

crop: area food collects before passing on to the gizzard

dewlap: loose skin on throat under beak

dubbing: removal of male gamefowl combs and wattles to prevent injury in fights

earlobes: area of bare skin below chicken's ear. Color denotes egg color, that is, in most cases white earlobes = white egg, red earlobes = brown or tinted

frizzle: a bird whose feathers curl in a random fashion, mostly toward the head

gizzard: the part of the chicken's digestive tract that contains grit to grind down the food

hackles: long, narrow feathers on bird's neck, also saddle feathers on male

hard feather: describes the short, narrow, tight-fitting feathers of a gamefowl

hen: female bird after its first molt

hen-feathered: a male bird that lacks sickles or hackle feathers, e.g., Campine

keel: the breastbone often featherless in gamefowl

leader: the backward-pointing spike of a rose comb

mandibles: upper and lower parts of the beak

meat spot: small, harmless spot of blood in the egg

molt: annual shedding of feathers, during which a hen stops laying

muff: feathers that protrude from both sides of the face in combination with a beard

parson's nose: lump of flesh from which tail feathers grow (technically known as uropygium)

point-of-lay: hens about to lay their first eggs, from 18 weeks old

pullet: female bird before its first molt

rooster: in theory used to describe a male after its first molt. In practice, it is used to describe a male chicken.

saddle: bird's back in front of the tail

self-color: single color all over

shrimp-tail: tail of some gamefowl, shaped like a shrimp

sickles: long, curved tail feathers of male birds

spike: see "leader"

spurs: sharp, horny growths on the legs of male and some female bird's legs; also known as "cockspurs" or "gaffe"

stag: male gamefowl before its first molt

tassel: see 'crest'

trio: group of two hens and one rooster

true bantam: small chicken that has no large equivalent

uropygium: parson's nose

vent or **cloaca:** orifice through which eggs and excretions are passed

vulture hocks: feathers that grow from the hock joint and point down

wattles: fleshy appendages that hang below beak

wing-clipping: when primary and secondary feathers are cut off or clipped on one wing to unbalance bird and prevent flight

Index

. . . .

Page numbers in *italics* refer to illustrations. Page numbers in **bold** refer to Glossary entries.